structure and form
in design

structure and form in design

critical ideas for creative practice

Michael Hann

London · New York

English edition

First published in 2012 by

Berg

Editorial offices:

50 Bedford Square, London, WC1B 3DP, UK

175 Fifth Avenue, New York, NY 10010, USA

Berg is an imprint of Bloomsbury Publishing Plc.

Library of Congress Cataloging-in-Publication Data

A catalogue record for this book is available from the Library of Congress.

British Library Cataloguing-in-Publication Data

A catalogue record for this book is available from the British Library.

ISBN   978 1 84788 743 6 (Cloth)

           978 1 84788 742 9 (Paper)

e-ISBN   978 0 85785 465 0 (individual)

Typeset by Apex CoVantage, LLC, Madison, WI, USA.

Printed in the UK by the MPG Books Group

www.bergpublishers.com

To Peter and Nell Hann

contents

acknowledgements

The author is indebted to Ian Moxon and David Holdcroft for their constructive review, useful commentary and helpful advice. Thanks are due also to Jeong Seon Sang (of Hanyang University, Korea) and to Alice Humphrey and Marjan Vazirian (both of the University of Leeds, United Kingdom) for their substantial efforts in producing large quantities of illustrative material. Gratitude is extended also to Briony Thomas, Christopher Hammond, Alison McKay, Robert Fathauer, Craig S. Kaplan, Josh Caudwell, Jeremy Hackney, Kevin Laycock, Ihab Hanafy, Dirk Huylebrouck, Jill Winder, Kate Wells, Margaret Chalmers, Hester du Plessis, Stephen Westland, Thomas Cassidy, Jae Ok Park, Young In Kim, Javier Barallo, Peter Byrne, Myung-Sook Han, Sookja Lim, Catherine Docherty, Behnam Pourdeyhimi, Francis and Patrick Gaffikin, Damian O'Neill, Brendan Boyle, Myung-Ja Park, Kyu-Hye Lee, Chil Soon Kim, Jin Goo Kim, Mary Brooks, Maureen Wayman, Kieran Hann, Charlotte Jirousek, Jalila Ozturk, Peter Speakman, Sangmoo Shin, Eun Hye Kim, Roisin and Tony Mason, Patricia Williams, Biranul Anas, T. Belford, Sandra Heffernan, Barbara Setsu Pickett, Eamonn Hann, Mairead O'Neill, Barney O'Neill, Ray Holland, Jim Large, Moira Large, Donald Crowe, Dorothy Washburn, Doris Schattschneider, Michael Dobb, Keum Hee Lee, Haesook Kwon, Young In Kim, Kholoud Batarfi and the BDP (the company of architects that designed the Victoria Quarter, Belfast), as well as the following student contributors: Zhou Rui, Alice Simpson, Natasha Purnell, Esther Oakley, Josef Murgatroyd, Robbie Macdondald, Natasha Lummes, Rachel Lee, Edward Jackson, Zanib Hussain, Jessica Dale, Daniel Fischer, Muneera Al Mohannadi, Nazeefa Ahmed, Elizabeth Holland, Olivia Judge, Alice Hargreaves, Reerang Song, Nathalie Ward, Matthew Brassington and Claira Ross. The author accepts responsibility for all omissions, inaccuracies and incorrect statements. Last, and by no means least, gratitude and thanks are extended to Naeema, Ellen-Ayesha and Haleema-Clare Hann. Every effort has been made to extend acknowledgement where it is due, and the author apologizes in advance should such acknowledgement be omitted. Unless specified otherwise, photographic images were produced by the author.

M. A. H

Leeds, 2012

list of illustrations

acknowledgements

Unless otherwise stated, all photographic work is by the author. Graphic and scanning work associated with geometric and other images is by Jeong Seon Sang (JSS), Alice Humphrey (AH), Marjan Vazirian (MV) and Josh Caudwell (JC), under the direction of the author.

figures

1

introduction

This book is concerned with structure and form in art and design. It covers a range of topics potentially valuable to students and practitioners engaged in any of the specialist decorative arts and design disciplines common in educational systems worldwide. The complexities of two-dimensional phenomena are explained and illustrated in some detail, and attention is paid also to explaining various three-dimensional forms. In the context of the decorative arts and design, structure is the underlying framework, and form is the resultant, visible, two- or three-dimensional outcome of the creative process. So, beneath the form of a design or other physical composition lies the structure. Structure includes grids, two- or three-dimensional lattices or other geometrical features which act as guidelines to determine the placing of components of a design. Such structural features may be hidden when the design is presented in resolved form or, occasionally, may be clearly visible when the design is complete. The hidden aspects of structure may include various systems of proportion such as the golden section and various root rectangles, as well as rules relating to geometrical symmetry. Both rules of proportion and rules of symmetry determine the placing of visual elements. The visible aspects of structure are those components used in the design's development which can still be seen when the design reaches a resolved, finished state. Examples include regular or irregular polygons, or other geometric figures forming a surface component that guides the placing of elements of a composition, pattern, image or other form of surface decoration (such as motifs held within squares following a strict geometrical order across a textile or other surface-pattern design), or the surface grid structures governing the design of the high-rise urban architecture of the twentieth and early-twenty-first centuries. Whether hidden or detectable visually in the final design, structure invariably determines whether or not a design is successful in terms of its aesthetics and its practical performance.

The series of preliminary actions associated with endeavours in the decorative arts, design and architecture more often than not involves the placing, in early development sketches, of one or more basic geometrical constructions, figures or combinations of geometrical elements. Further, various systems of proportion appear to have been applied through the centuries as underlying features in the design of certain buildings, structures, crafted and manufactured objects and surface-pattern decorations. In ancient times knowledge of the rules governing the use of such systems of proportion may well have been passed from skilled craftspeople to apprentices, sons or daughters. In more recent times such knowledge appears to have been largely lost, forgotten or ignored by many (though by no means all) creative practitioners, and decisions relating to composition or positioning of components or elements of a design have, in many cases, been guided by instinct, past experience or an educated subjectivity, possibly involving also elements of guesswork (or try it and see). In many areas of creative endeavour, such approaches may lead to satisfactory results. However, it may be that a more structured approach, based on an understanding of the range of concepts mentioned above, together with various related guidelines, can inspire confidence at the early stages of the design process and ensure that a satisfactory design solution is presented, possibly with greater efficiency. For this reason, substantial emphasis is placed in this book on identifying various geometric concepts and in presenting and discussing a range of simple guidelines to assist the creative endeavours of both accomplished and student practitioners, teachers and researchers. While the focus is on explaining concepts and principles associated largely with two-dimensional space, the applicability or relevance to the realms of three dimensions, where appropriate, is noted also. The intention is not to offer guidelines on precisely how a designer should engage with the complete design process and how the various stages of problem solving should be approached, but rather to present a foundation which should act as a knowledge framework to guide the designer in the early stages of addressing a design brief. While the student designer is the principal intended audience, there is much of value for experienced (and not so experienced) practitioners as well as design researchers and other observers who wish to explore the fundamentals governing structure and form in art and design.

Mathematical analysis and procedures have been used by various researchers to detect the use of systems of proportion and other geometrical features in designed objects (including various product designs, buildings and other constructions). The author maintains that knowledge of these systems of proportion and related features is of value to designers in the twenty-first century. An appreciation of how they can be applied in the context of the visual arts and design does not depend, however, on advanced mathematical background knowledge, but simply on an awareness of certain basic geometrical constructions of the type covered early in high school. The reader is referred to the useful primer *Drawing Geometry* by Jon Allen (2007), should a review of basic geometrical procedures and principles be required. The intention of this current book is to provide students and creative practitioners with an appreciation of applicable structures, procedures and methods which does not rely on a substantial understanding of mathematical principles or a comprehension of unfamiliar terminology, symbols or formulae.

Certain general organizational principles or considerations have been associated by various commentators with successful designs. Among the most commonly listed are balance, contrast (of tone, shape, colour and texture), rhythm, form following function, the 80/20 rule, the rule of thirds, the golden section, iteration, modularity, ground-and-figure relationships, and various principles of perception and visual organization drawn from the realms of *Gestalt* psychology (including proximity, similarity, continuance, closure and *Prägnanz*). These are all of varying importance depending on the specific intention, end use and message to be conveyed. Many (though by no means all) of these concepts enter the explanations and discussions in the following chapters (which, as stated previously, are focused primarily on structure and form). Readers wishing to develop a more complete understanding of the fuller spectrum of issues are advised to consult the following texts: Lupton and Phillips (2008) and Lidwell, Holden and Butler (2003).

The book is arranged as follows. Chapter 2 focuses particular attention on the two fundamental structural elements of point and line, for it is these which make all structural arrangements and resultant forms possible. Chapter 3 identifies, explains and illustrates a selection of geometrical constructions traceable to ancient times and usable in the decorative arts, design and architecture. Chapter 4 introduces and illustrates various categories of periodic tiling and tessellation, and also explains the nature of Penrose-type tilings. Chapter 5 is concerned with geometrical symmetry and shows how the concept can be used to characterize types of regularly repeating frieze and all-over patterns. Concepts associated with scale symmetry and fractals are introduced also. Chapter 6 explains the Fibonacci numerical series and introduces constructions such as the golden section and related rectangles and spiral forms. Chapter 7 is concerned with various three-dimensional phenomena, including polyhedra, spheres and dome structures. Chapter 8 focuses primarily on a selection of concepts and issues of importance to structure and form in three-dimensional design. Chapter 9 reviews the characteristics of modularity across a range of areas, and highlights the relationship between modularity, closest packing and efficient partitioning. Chapter 10 considers structural analysis in the decorative arts, design and architecture, and presents steps towards the development of a systematic analytical framework aimed at providing a consistent method for future design analysts. Chapter 11

presents a summary of the principal components of the book and, in particular, identifies those procedures, methods and approaches which may prove of value to practitioners involved in the decorative arts, design and architecture. The bulk of illustrative material was prepared under the direction of the author. Illustrative examples of work from a selection of independent theorists, artists and designers are included at various points in the book. A substantial proportion of illustrative material is included also from art-and-design students registered on courses in design theory presented by the author at the University of Leeds (United Kingdom), Hanyang University (Korea) and Yonsei University (Korea). Various sample exercises and assignments are included in Appendix 1.

It is worth remarking at this stage that it is assumed often that geometry, when used in art and design, will necessarily result in stiff geometric, hard-edged, sharp-lined solutions, and that more natural, free-flowing forms only result from some sort of spontaneity without predetermined underlying structure. This is plainly not the case. Japanese paper stencils (known as katagami) are renowned as expressions of free-flowing forms, but it should be noted that these are underpinned largely with a strict geometric structure aimed at attaining a carefully measured balance brought about by considerations relating to symmetry and asymmetry. Katagami stencils result from careful planning and delicate execution, and are used in resist-dyeing of textiles destined for various forms of traditional Japanese garments. The process of stencil cutting is highly skilled and involves cutting designs into sheets of laminated mulberry paper, often reinforced with grids of silk thread; some claim that this is the precursor to silk-screen printing of textiles. Japanese paper stencils were a great influence upon the work of many artists and designers, including Vincent Van Gogh, Frank Lloyd Wright, James Mc Neill Whistler and Louis Tiffany. A series of illustrations of katagami stencils is presented here (Figures 1.1–1.12). It is worth noting that although the designs presented in many cases are apparently free-flowing, all are underpinned with a strict adherence to various geometrical rules, regulations and procedures, many of which are the focus of attention in this book.

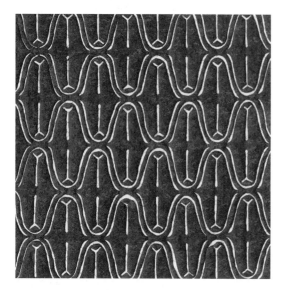

Figure 1.1 Katagami stencil 1, 'Fishing net 1', Courtesy of University of Leeds International Textiles Archive.

Figure 1.3 Katagami stencil 3, 'Gems 1', Courtesy of University of Leeds International Textiles Archive.

Figure 1.2 Katagami stencil 2, 'Cranes over water', Courtesy of University of Leeds International Textiles Archive.

Figure 1.4 Katagami stencil 4, 'Pine needles and pine cones', Courtesy of University of Leeds International Textiles Archive.

Figure 1.5 Katagami stencil 5, 'Gems 2', Courtesy of University of Leeds International Textiles Archive.

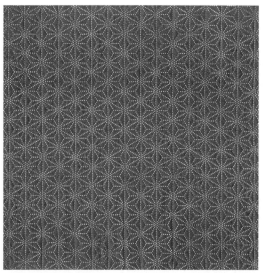

Figure 1.7 Katagami stencil 7, 'Hemp leaves', Courtesy of University of Leeds International Textiles Archive.

Figure 1.6 Katagami stencil 6, 'Fishing net 2', Courtesy of University of Leeds International Textiles Archive.

Figure 1.8 Katagami stencil 8, 'Wisteria and stems', Courtesy of University of Leeds International Textiles Archive.

Figure 1.9 Katagami stencil 9, 'Ornamental vine', Courtesy of University of Leeds International Textiles Archive.

Figure 1.11 Katagami stencil 11, 'Lobsters and waves', Courtesy of University of Leeds International Textiles Archive.

Figure 1.10 Katagami stencil 10, 'Thunder, lightning and clouds', Courtesy of University of Leeds International Textiles Archive.

Figure 1.12 Katagami stencil 12, 'Hairy-tailed turtles', Courtesy of University of Leeds International Textiles Archive.

the fundamentals and their role in design

introduction

Various fundamental elements play a crucial role both in the visual arts and in design. Depending on what is being designed, the elements of point, line, shape, scale, plane, texture, space, motion, volume or colour may be of greater or lesser significance. Elements such as these were considered by Dondis to constitute 'the basic substance of what we see', and could, in her view, be brought together to provide contrast at one extreme and harmony at the other, as well as at the stages between these extremes (1973: 39). Among the extremes (or 'techniques') listed by Dondis were instability and balance; asymmetry and symmetry; irregularity and regularity; complexity and simplicity; fragmentation and unity; transparency and opacity; variation and consistency; distortion and accuracy; depth and flatness; sharpness and diffusion (1973: 16). The nature of these and related elements and principles together with their role in visual communication has been covered in various basic works, though not surprisingly different publications give different lists of elements. Arnheim (1954 and 1974) presented a comprehensive discussion, cast in psychological terminology, of the nature of a range of visual elements or principles, including balance, shape, form, growth, space, light, colour and movement.

Lidwell, Holden and Butler (2003) showed how, together with other elements, point and line were paramount considerations to designers, and explained how various principles influenced the perception of designs, how they communicated information and how designers could improve the usability and appeal of designs. More recently, Lupton and Phillips (2008) considered point, line, plane, rhythm, balance, scale, texture, colour, figure/ground relationships, framing, hierarchy, layers, transparency, modularity, grids, patterns, time and motion, rules and randomness, and showed how these were important considerations to designers.

Scholarly understanding of fundamental issues relating to structure and form in architecture and design has been influenced deeply by the monumental contributions associated with the Bauhaus. A cross section of useful commentary, sources and reviews includes Kandinsky (1914 and 1926), Klee (1953), Schlemmer (1971, Eng. ed.), Tower (1981), Naylor (1985), Lupton and Abbot Miller (1993), Rowland (1997) and Baumann (2007). Other important twentieth-century texts include Le Corbusier (1954), Wolchonok (1959), Itten (1963), Critchlow (1969), Wong (1972 and 1977), Pearce (1990), Kappraff (1991) and Ching (1996). It is not the intention of this book to cover again all the material dealt with by these and other texts, but rather to identify specific aspects of structure and form which are of value when applied in the visual arts and design.

This chapter focuses particular attention on the two structural elements of point and line, for it is from these two elements that much else follows (Figure 2.1). The perspective taken is in sympathy with that of Kandinsky (1926), who recognized that an awareness of the nature of point and line was fundamental to understanding the nature of structure and form, the principal focus of this book. A further intention of this chapter is to introduce various simple polygons, and also to explain the nature of grids and their organizational value to designers.

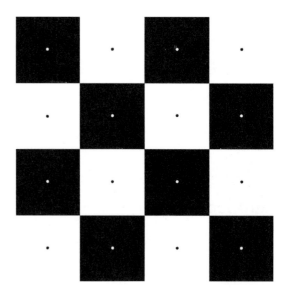

Figure 2.1 'Point, line and plane', by Ihab Hanafy, 2011.

point—a sound of silence

Design is a process of visual enquiry which takes into account aesthetic and practical considerations. Two particular structural elements, point and line, are considered to be vital building blocks governing and assisting this process. Both can be detected in abundance in our environment, in both modern-day and ancient structures (Figures 2.1–2.7).

Both point and line can be considered to have internal meaning and significance on the one hand as well as external function and application on the other. Both the internal and external aspects are of importance to the designer. In formal geometric terms (the internal aspect), a point is without dimensions and denotes a position in space. It is a place without an outside or inside and cannot be measured because there is nothing to measure. Its position may be indicated by the crossing of two lines, and recorded or denoted by reference to coordinates of two or maybe three axes. When

considered in terms of physical content, it equals zero. In terms of duration of time, the point exists within an instant. It is static and silent, and this metaphorical silence is a dominant feature. Kandinsky commented that 'the sound of that silence customarily connected with the point is so emphatic that it overshadows the other characteristics' (1979: 25). Despite this silence, the point is the source from which all geometry springs and acts as a positional marker in the vast bulk of geometric constructions. This role as positional marker is of particular importance in the context of design and construction as points may indicate the positions or locations of aesthetic or practical features underpinning the design's structure.

So if a point is dimensionless and barely, if at all, visible, of what benefit can it possibly be to designers? Designers engage in a problem-solving process dependent invariably on visual representation which goes through successive stages until the process is complete. Visual representation may involve, for example, ink and paper or pixels

Figure 2.2 'Points around 1', Seoul, 2010.

Figure 2.3 'Points around 2', Seoul, 2010.

on a screen. In order for such visual representation to progress, it is conventional practice to give the point usable dimensions, so a tiny, circular form (or dot) is invariably employed; this is the external aspect, and this is the focus of the further discussion that follows.

An instance where the point is given physical form is in language and writing; as the full stop or period, it denotes an interruption, a gap or bridge between one statement and another. It separates the past from the future and can thus be considered in the present. Often, the point is perceived as round and small. Both shape and size are relative. The outer limit of size is determined by the relative size of the plane around it, and also the relative sizes of the other forms in the

Figure 2.4 'Lines around 1'
Seoul, 2010.

Figure 2.5 'Building lines 1', Seoul, 2010.

Figure 2.6 'Building lines 2', Manchester, 2011.

Figure 2.7 'A ship from years past', 2011, photo courtesy of Jeremy Hackney.

same plane. As regards shape, the possibilities are seemingly limitless: a triangular, oval, square or any other regular or irregular shape is feasible, though a simple circular format is most common. Of course, externally, the full stop is merely a sign with a practical purpose.

In architecture and other three-dimensional design, a point may result from the termination of an angle in space (e.g. spires on Gothic cathedrals) or where two or more surfaces (planes) meet at an angle (Figure 2.8). The tips of an arrow, dart, thorn, or pen nib are other examples. Graphically, a point gains significance when placed within a frame such as a square. When positioned at the centre, a point is at ease and holds attention as a focal point; when a point is shifted off centre, the sense of repose is upset, and a visual tension arises. Points are used in maps and plans to represent towers, spires, obelisks, and other locations of interest on the landscape.

A point, therefore, has no content, weight, dimensions, outside or inside. On its own, it has no meaning other than its position. When given visible dimensions (in order for it to be seen) and when placed in association with another point in the same plane, the tendency of the observer is to connect the two (in the mind's eye) so that an apparent or imaginary line is created. Likewise, when three points which cannot be connected by a straight line are placed on the same plane, the tendency is for the mind's eye to create a triangle. Lines can of course be real (in that they can be represented in the plane by a continuous mark made on paper, for example, by pencil or ink) as well as imagined.

line and what follows

So, from the dimensionless point comes the line which, in geometric terms (the internal aspect), can be considered either as a moving point or as the path between two points. In the former case, a line results from external force acting on a point in order to move it in a specified direction. A line can therefore be considered as the result of a leap from one state to another. Alternatively, when considered to be lying between two points (with one point at each extremity), the line is in a state of continuous energy. While the single point is a static entity, the line can be considered as a force between two points and is thus in a dynamic state.

A conceptual line has a beginning and an end, has length but no width, and is thus one-dimensional. Again, as with the point, in order for this element to be of value to designers, it needs to take on a substantive form (the external aspect). In practice, a line is given width in order to make it visible (e.g. a pencil line on paper), and many different widths are possible (until common sense suggests that measurable breadth has been created). Lines may be of infinite lengths, may have a wide variety of weights (e.g. emboldened or subtlety shaded) and may be straight (the minimum distance between two points) continuous, discontinuous (though the perception of continuity is retained), wavy (with curved sections) or angular (acute, e.g. 60 degrees; right-angled, i.e. 90 degrees; or obtuse, e.g. 120 degrees). Lines have psychological impact, which will be influenced by their direction or orientation, weight and emphasis, and variations in these. Lines may be created by nature or may be human-made. They may exist by implication, at an interface between two colours or textures, for example. Lines may be used by the visual artist to express equilibrium if depicted horizontally in the plane (any flat surface), stability if depicted vertically in the plane and activity or movement if depicted diagonally in the plane. Together, lines are used in the design-planning stage to indicate the boundaries of shapes, areas or masses which define forms in two- or three-dimensional space. Line is the essential ingredient

used by all designers, architects and visual artists, and is the means by which the imagined is represented in visual form. Lines can be given energy, can be made dynamic and can be given purpose and focus. When used to sketch an idea, lines may be loose and free (as with a quick pencil sketch of a still life, for example) or tight and measured (as in a finished engineering drawing).

Rods, columns and scaffolding are linear in nature and, if of sufficient strength, can provide structural support in buildings, bridges and numerous other designed objects and constructions. In a completed design or construction, lines may not be detectable visually as clear linear devices, but rather may be implied and exist only as the interface between two or more types of

Figure 2.8 'Point, with line and plane', Manchester, 2011.

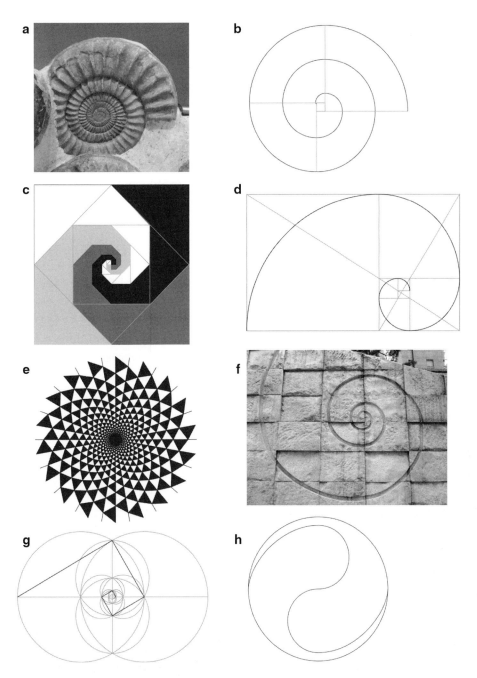

Figure 2.9a–h Spirals of various kinds, image prepared by AH.

surface, colour, tone, plane or texture. In much architectural (Figure 2.8) and product design, lines are the meeting points of two planes or alternatively constitute the outline of the object or construction. Line dancing is where groups of people move in lines during the course of a dance. Lines in nature, tracks in snow, railway lines, sky lines, silhouettes, tree branches in winter, justified text and columns of type on a page, railings and linear perspective lines are some additional examples.

A spiral is a continuous line with a regularly changing orientation. Spiral-type growth in the natural world is found in ram's horns, the nautilus shell, the common garden snail's shell, sunflowers, pineapples and pine cones. The spiral has been used in architecture since the time of the ancient Greeks and has been employed as a decorative element in numerous cultures for several millennia. Extensive treatises dealing with spiral formations, mainly in the natural world, have been provided by Davis (1993) and Cook (1914). Williams (2000) provided an interesting short article on spirals and rosette motifs in architectural ornament (mainly Greek, ancient Roman and Italian Renaissance). The spiral is found in both the natural and constructed worlds and there are several varieties from a geometrical viewpoint (Figure 2.9a–h).

planes, shapes and forms

Both point and line are most easily considered as lying on a plane. In fact, all geometric figures are invariably represented on the plane (for example, drawn using pencil on paper). Described geometrically as a moving line, a plane is a flat surface with length and breadth, and may be solid or perforated, monocoloured, multicoloured, transparent or opaque. Examples include this page, a page from a sketch book, most canvasses, most sheets of paper and a computer screen. Using point and line in the plane, applied by pencil, pen, brush or charcoal, designers and artists can create representations of landscapes, faces, objects, patterns, typographic statements,

posters or other types of visual communication, architectural and engineering drawings and charts, which communicate all sorts of quantitative and qualitative data.

Point and line can be seen to relate closely also to structure and form, both of which are invariably represented by the designer using the two-dimensional plane. Collectively, points and lines can be used to represent the structure of a two- or three-dimensional form. Structure is thus best considered as the (invariably) hidden framework or skeleton underpinning form (e.g. branches of a tree in full leaf, steel girders of a building, the geometric grid underlying a regularly repeating pattern). The designer can represent both two- and three-dimensional forms on the plane although three-dimensional forms, in reality, invariably require the meeting or combination of more than one (flat) plane.

Shape is the external appearance of a design (or a component part), defined by the outline whether in two or three dimensions. Examples include the silhouette of a dress or the outline of a chair, a kettle, a surface-pattern motif, an automobile or a building. Designs in two or three dimensions, when rendered on a flat surface, may be composed of a collection or arrangement of two-dimensional shapes. Shape is the '...characteristic outline or surface configuration of a figure or form' (Ching 1998: 23). Arnheim observed that 'the physical shape of an object is determined by its boundaries–the rectangular edge of a piece of paper, the two surfaces delimiting the sides and the bottom of a cone' (1974: 47).

A shape is always considered in the context of the area or space surrounding it. Ching commented: 'A shape can never exist alone. It can only be seen in relation to other shapes or the space surrounding it' (1998: 23). In a two-dimensional representation of a series of objects, the background area of the drawing is considered as negative space and the foreground shapes as positive. When viewing all such representations, we generally ignore the background negative shapes and focus instead on the shapes we consider to be positive, although negative and positive shapes share the same contour lines. Contour lines present outlines of the component shapes of an object and indicate on a two-dimensional surface the edges between different planes, tonal values or textures of that object. A good explanatory discussion of contour drawing was given by Ching (1998: 17–22).

In the monumental treatise *On Growth and Form,* first published in 1917, D'Arcy Wentworth Thompson observed that the form of an object has an underlying 'diagram of forces' which symbolizes or represents manifestations of various kinds of energy, including the forces present when the object was created (1966: 11). Although

focusing on the biological context, Thompson's work is of immense relevance in the context of structure and form in design. The idea that form is underpinned by a 'diagram of forces' was developed further by Pearce, in his consideration of the principle of 'minimum inventory/maximum diversity' in the context of architectural design (Pearce 1990: xii).

various polygons

A polygon is a closed figure bounded by straight lines (or sides). A triangle is a polygon with three sides and three interior angles (each the meeting point where two sides meet). In the context of design, the most commonly used triangle types are the isosceles triangle, with two sides and two angles equal; the equilateral triangle, with three sides and three angles equal; the right-angled triangle, with one angle equal to 90 degrees

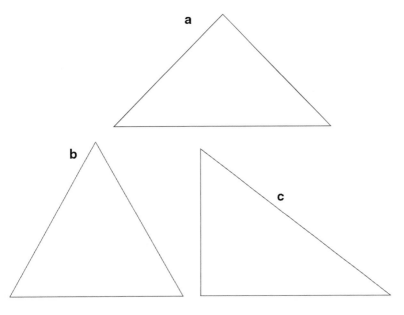

Figure 2.10a–c Three triangles (JSS).

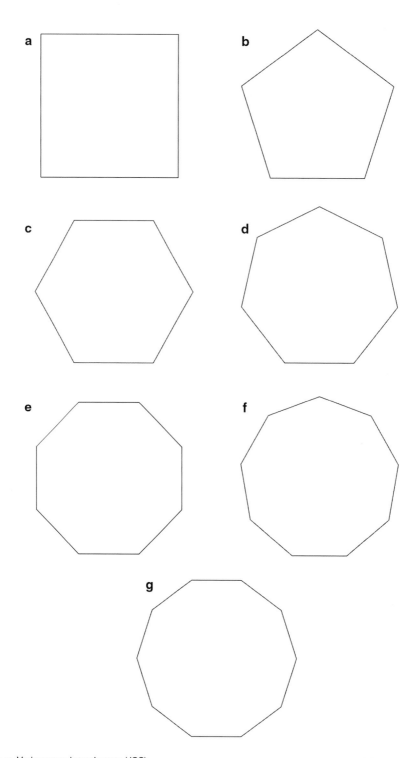

Figure 2.11a–g Various regular polygons (JSS).

(Figure 2.10a–c). A regular polygon has equal-length sides (equilateral) and equal angles (equiangular). A square has four equal-length sides and equal angles, a regular pentagon has five equal-length sides and equal angles, a regular hexagon has six equal-length sides and equal angles, a regular heptagon has seven equal-length sides and equal angles, a regular octagon has eight equal-length sides and equal angles, a regular enneagon has nine equal-length sides and equal angles and a regular decagon has ten equal-length sides and equal angles (Figures 2.11a–g).

As observed by Ghyka (1977: 22), the most important non-equilateral triangle is the right-angled triangle with sides of three, four and five units (Figure 2.12). The 3-4-5 triangle, also known to Vitruvius (the engineer and architect of the emperor Augustus, and author of a series of important books) as the Egyptian triangle, was of great significance in ancient times and was used seemingly, in the construction of various of the Egyptian pyramids (Kappraff 2002: 181). It is associated with the *harpedonaptae* (or rope stretchers), seemingly the first land surveyors or engineers, who were sent out when the annual flood waters of the Nile had slipped back to redraw the land markers of fields suitable for cultivation so that taxes payable to the Pharaoh could be assessed fairly. This task was achieved through the use of a measuring stick of some kind plus a rope with twelve equal divisions (suited to create a right-angled triangle of sides three, four, and five). Subsequently, the 3-4-5 triangle was used by the ancient Greeks. According to Ghyka such a triangle was called the 'sacred triangle of Pythagoras, or of Plutarch', and it was used also by ancient Persian (Achaemenid and Sassanian) architects (Ghyka 1977: 22).

A third important triangle from the viewpoint of structure and form in art and design is the

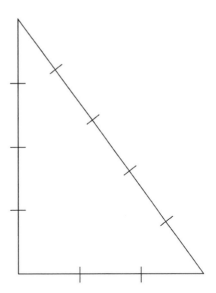

Figure 2.12 A 5-4-3 triangle (JSS).

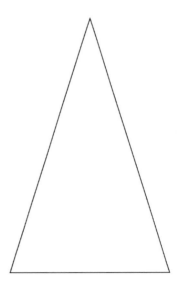

Figure 2.13 36-degree isosceles triangle (JSS).

isosceles triangle having 36 degrees at its smallest angle (Figure 2.13). Ghyka (1977: 23) referred to this as the 'sublime triangle' or 'triangle of Pentalpha', presumably because of its potential use in creating a regular pentagram (a five-pointed star).

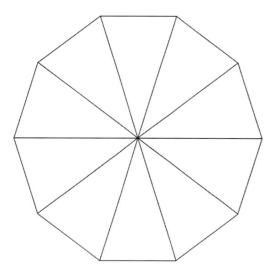

Figure 2.14 Ten 36-degree isosceles triangles (JSS).

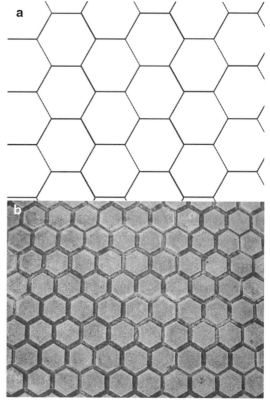

Figure 2.16a–b Regular hexagonal grid (JSS).

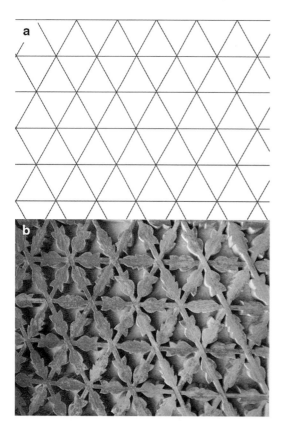

Figure 2.15a–b An equilateral-triangle grid (JSS).

It is worth noting also that ten of these triangles can fit exactly together as shown to create a regular ten-sided figure (or decagon). This is readily achievable because ten of the small angles, each of 36 degrees, fit exactly into 360 degrees (Figure 2.14).

Certain regular polygons combine to produce various tiling designs. In the context of this book, a tiling is an assembly of polygon shapes which cover the plane without gap or overlap (a topic dealt with in detail in Chapter 4). The three basic types of tilings consist of component cells of three distinct regular polygons: the equilateral triangle

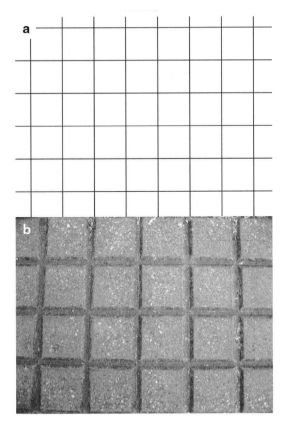

Figure 2.17a–b Square grid (JSS).

(Figure 2.15a–b), the hexagon (Figure 2.16a–b) and, most commonly, the square (Figure 2.17a–b).

grids and guidelines

A grid is an assembly of lines, typically evenly spaced and with half running horizontally and at right angles to the other half which runs vertically across the plane. Grids have played a major role in design and architecture at least since the beginning of the industrial age (though there is much evidence for their use before). The value of grids is not simply to transpose impressions of three-dimensional objects on to a drawing surface (as did some European artists during the sixteenth century), but rather as organizational structures to help in the determination of the exact relative locations of components of a design. Grids offer a basis for achieving a harmonious relationship between the constituent visual elements of a design or other visual composition, and are thus of great potential value to artists and designers.

Grids may be based on some particular geometrical construction which in turn may yield special ratios or proportions or may, at the points of intersection of lines, indicate the ideal positioning of key components of a design or other composition. Correct positioning can impart balance to a visual composition. Balance holds the components together and is often created simply through reflectional symmetry with identically shaped and sized parts placed on the grid, equidistant on either side of an axis (or an imaginary two-sided mirror). Balance can be achieved also by asymmetrical positioning of elements in a composition through arrangements on the grid similar to a see-saw, with heavier weights positioned more closely to the fulcrum balanced by lighter weights positioned farther away from the fulcrum.

Grids provide structural frameworks to guide the development of designs and the placement of component parts. Just as the horizontally oriented lines on typical notepaper act as guidelines to ensure horizontal orientation of handwriting, grids can be useful organizational tools in, for example, the drawing of floor plans and the layout of landscapes, building façades and engineering drawings. Various grids are of value in organizing repeat units in patterns and tiling designs. In the design of woven textiles, squared paper (known as point paper and arranged in adjacent blocks of eight by eight) is used to indicate the desired interlacement of threads (Figure 2.18a–b). Often, such renditions serve the intermediate stages of activity between the designer and the producer. Indeed, many highly skilled hand-knotted-carpet weavers (e.g. among the Bedouin in the first decade of the twenty-first century) could only follow the time-honoured interpretation of a design shown in squared-paper form, together with associated symbols indicating scale, colours and types of yarn to be used, and were unable to respond positively to a fully coloured, precisely rendered drawing of the intended outcome. The weaving of highly figured textiles (using a Jacquard loom) throughout much of Europe during and before the early-twentieth century relied on several stages between designer and weaver; invariably, the weaver would be unable to respond to a fully coloured design, but rather required the design to be rationalized initially into grid or squared-paper form and then transferred to a series of punched cards (with which to programme the loom). Similarly, knitting point paper (which has a lattice of points) indicates the anticipated disposition of loops of yarn on needles during the knitting process. In typographical organization, in years past, a grid acted as a compositional device to ensure aligned columns of text, precise margins and consistent page templates.

As well as acting as a structural framework to guide the development of a design, a grid can occasionally be the resultant visual feature of a finished design. A commonly cited example from the late-twentieth century is high-rise architecture

Figure 2.18a–b Point-paper design and resultant woven fabric.

Figure 2.19 High rise in Seoul, 2010.

which invariably exhibits grid structures of steel, concrete and glass (Figure 2.19). Seen from afar, cultivated fields on the landscape exhibit grid structures: invariably, these are irregular, particularly where small-holdings predominate; occasionally, where private combines or state ownership are predominant features, relatively large-scale areas are set aside for cultivation, and large rectangles or squares form the cells of regular landscape grids.

Regular grids from given-sized squares, equi-lateral triangles or hexagons are commonly used in the construction of regular repeating tilings and patterns (dealt with in Chapters 4 and 5, re-spectively). Several other grid types, constructed through the use of certain ratios or proportions, are of value to designers at the planning stage. So from a standard 1:1 grid, others in whole number ratios such as 1:2, 2:3, or 3:4 (Figures 2.20a–d) may be produced. Alternatively, grids may be based on certain ratios such as 1.618:1, a ratio found commonly in nature and, according

to numerous scholars, used by artists, designers and architects for the past two millennia or more (this ratio is discussed in Chapter 6).

Designers should also be aware of grids based on the DIN system of paper sizing. Such grids do not repeat in the same way as those based on the three regular polygons mentioned previously, but rather offer an area which may act as a (sym-metrical or asymmetrical) framework within which to place key components of a design or other composition. These key components could, for example, be text and images in the case of Web or poster design; paving, pathways, cultivated areas, walls or terraces in the case of landscape design; various building features in the case of ar-chitectural design; areas of pattern, texture, co-lour, light or shading in the case of carpet or other surface-pattern design. A worthwhile exercise is to take a standard size of paper (such as A4) and divide it into four equal parts (Figure 2.21a) and subsequently divide these parts using the ratio 1:1.414 (which, with A4-sized paper, is the ratio

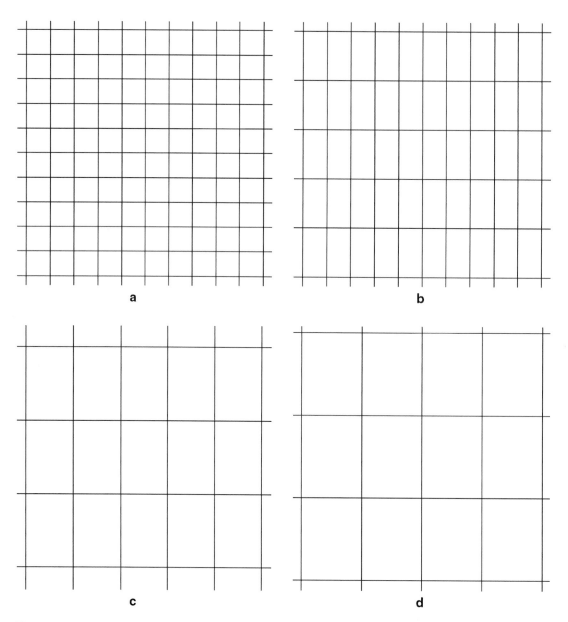

Figure 2.20a–d Grid developments (JSS).

of the length of one short side to the length of the long side), laying each successive division on the previous (Figures 2.21b–c show two successive stages). The resultant rectangular cells are not all equal in size. At the same time, they share

a harmonious relationship and exhibit geometric complementarity (that is, though of different actual sizes, the rectangular cells exhibit a harmonious relationship with each other). The phrase *geometric complementarity* has been coined to

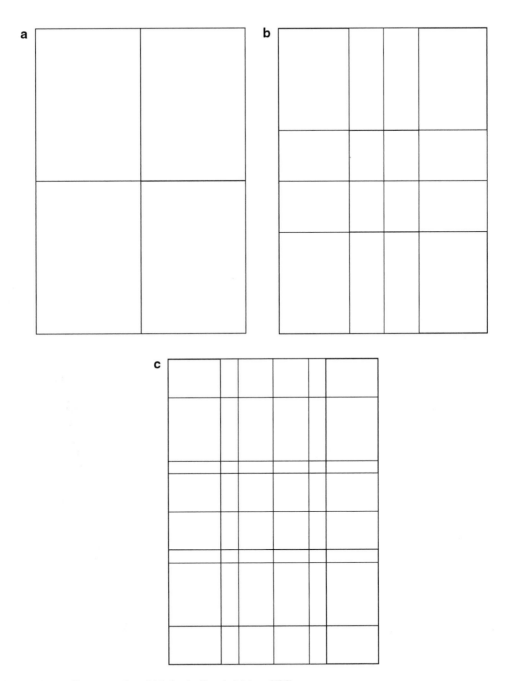

Figure 2.21a–c Representation of A4 sheet with subdivisions (JSS).

refer to the tendency for geometrical figures, from the same parent source but of differing dimensions, to work in visual harmony with each other, especially if they share the same proportion. In the case described earlier, the divisions of a standard-sized piece of paper offer a template to create a balanced, harmonious composition by providing areas to place key visual components of the composition. Further varieties are introduced later in association with various important constructions considered next.

The rule of thirds is an editing or compositional technique or guideline which creates a grid by dividing the design field (invariably the two-dimensional plane) into thirds both horizontally and vertically, thus creating a grid of nine rectangles and four intersections. These intersections are often (though not always) useful as aesthetic points at which to locate key visual features or practical components of the design. It is believed that the technique was used by Italian Renaissance painters as a compositional tool. In modern times it is commonly used in press photography, poster and Web-page design (Lidwell, Holden and Butler 2003: 168).

When a grid is used as a compositional tool or as an indicator of the disposition of design elements, physical features, colours or textures, the scale of the design drawn in relation to the manufactured, built or created object is a crucial consideration. So, at the beginning of the design process the designer must decide whether one cell of the grid equals 10 centimetres, 1 metre or 50 metres of the design when it is realized in physical form. Scale is of course a relative concept. An object, shape or form can only be deemed large in scale when considered in relation to a similar object, shape or form which, by comparison, is deemed to be small in scale. A designer works invariably with a specified scale of application in mind, although some design briefs demand a result equally applicable at various scales: handheld (e.g. phone screen or magazine page), poster-size (say of 1-metre height), or large-scale (e.g. cinema screen or billboard). Within a

composition, contrast in scale can hold the attention of the viewer and can give an impression of distance: smaller shapes appear to recede, and larger shapes seem to come to the foreground. Road maps have a specified scale (e.g. 1 centimetre to 1 kilometre, where 1 centimetre on the map represents 1 kilometre of road). The scale of the human body is a crucial reference point for the majority of designers (especially those involved with product, textile, fashion, interior and architectural design). If scale is not accounted for early in the design process, despite the fact that the design may appear to be well resolved on the computer screen or paper grid, the resultant design may not be suited to the intended end use. Children's artwork often offers a confused understanding of the rules of linear perspective (Figure 2.22).

use to the creative practitioner

Regular grids can undergo transformations, and various angled grids as well as curved grids are possible (Figures 2.23a–d). The scale of successive cells can be diminished or enlarged and their orientation adjusted in a systematic way (e.g. Figure 2.24). By using commonly available software, various further manipulations are possible (Figure 2.25 a–k).

Several systematic variations based on the regular square grid can be proposed. The balance between vertical and horizontal can be adjusted in one direction or both by changing the individual unit to a rectangle (Figure 2.26a–d). This is a theme that is taken up further in a later chapter when attention is given to various specially proportioned rectangles. The orientation and angles of lines can be changed to produce diamond-shaped

Figure 2.22 Unknown child artist, Seoul Station, Korea, 2008.

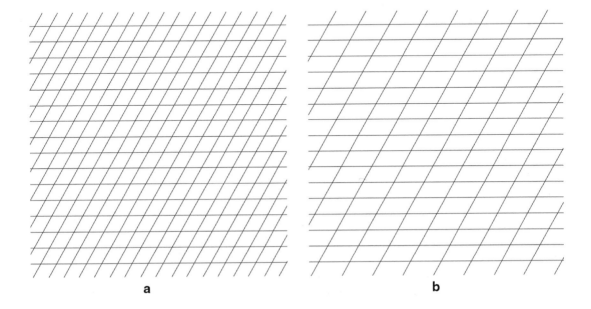

a

b

Figure 2.23a–d Various grids (JSS) (continued).

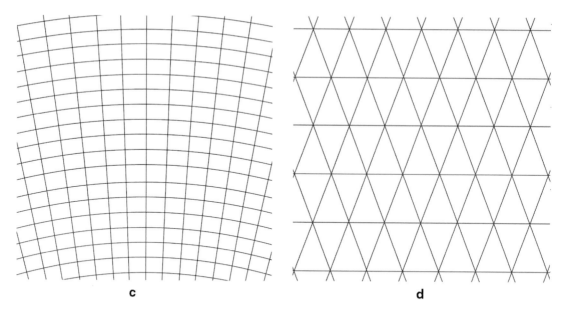

Figure 2.23a–d Various grids (JSS).

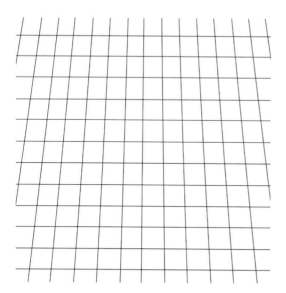

Figure 2.24 Grid with slight scale change (JSS).

variations (Figure 2.27a–d). Rows or columns of cells can slide from left to right or up to down respectively, systematically involving all rows or columns or alternate rows or columns (examples given in Figure 2.28a–b). The component vertical, horizontal or both sets of lines can be curved or angled in a regular fashion to ensure series of cells of the same size and shape (Figure 2.29a–d). Through systematic manipulation of the cells in a simple square grid, by combination or removal as well as simple colour variation, numerous additional varieties are possible. Variations of all of these techniques can be applied to the other two commonly used regular grids (with unit cells of equilateral triangles or regular hexagons) to provide equally extensive variations.

From the viewpoint of designers wishing to generate series of regular repeating patterns, it is particularly beneficial to work from grid structures, and there are of course many varieties of these. As will be see in Chapter 4, there are numerous tiling designs which can function equally as well as grids and which offer the ideal platform to the designer wishing to generate regular-repeating-pattern ideas.

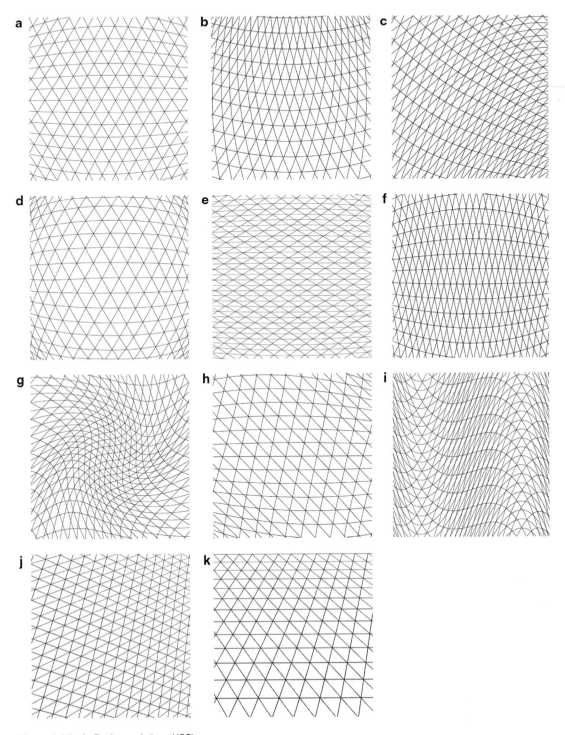

Figure 2.25a–k Further variations (JSS).

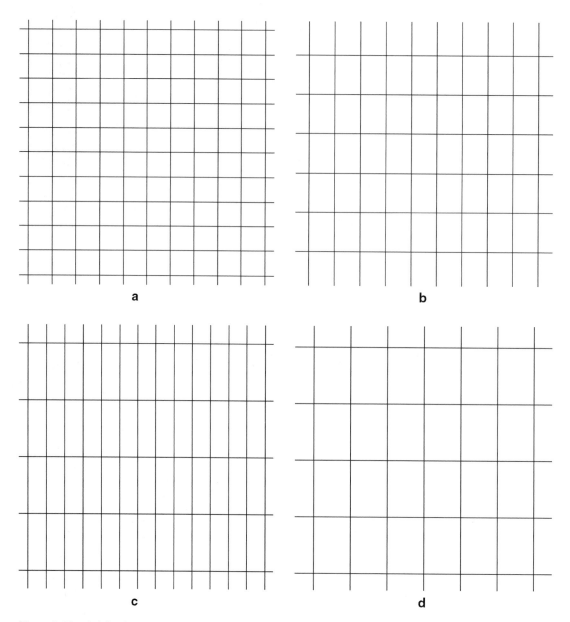

a

b

c

d

Figure 2.26a–d Adjusting cell size (JSS).

Figure 2.27a–d Adjusting orientation (JSS).

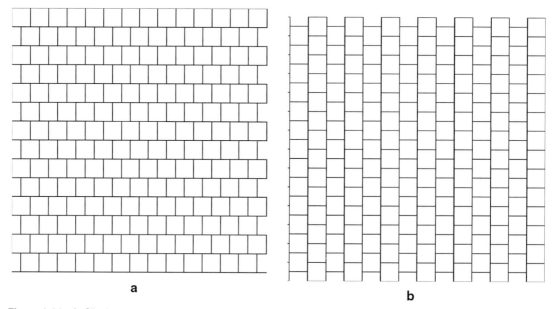

Figure 2.28a–b Slipping across and slipping down (JSS).

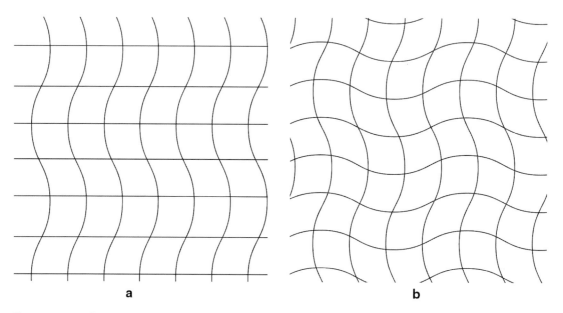

Figure 2.29a–d Curved or angled (JSS) (continued).

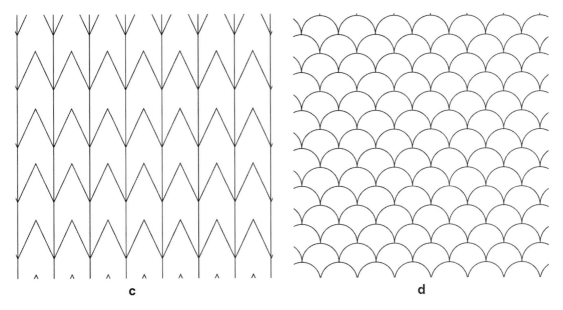

Figure 2.29a–d Curved or angled (JSS) (continued from previous page).

CHAPTER SUMMARY

This chapter has focused attention initially on the two fundamental structural elements used by artists and designers: point and line. When considered in strict geometrical terms, the point is dimensionless and denotes only a location, while the line has only one dimension and is best regarded as connecting two points. In order for both points and lines to be of value to artists and designers, they need to be given more substantive presence. Collectively, points and lines can be used to represent structures and forms. In the context of this book, structure is considered as the often hidden framework underpinning form, whether in two- or three-dimensional space (though in each case invariably represented on the two-dimensional plane, possibly by using pencil on a sheet of paper). A range of closed geometrical figures was introduced, including equilateral and other triangles, and regular polygons, including the square, pentagon and hexagon. The phrase geometric complementarity was coined to refer to the tendency for geometrical figures, from the same parent source but of differing dimensions, to work in visual harmony with each other, especially if they share the same proportions. A range of regular grids was introduced and a selection of possible systematic variations and transformations suggested. Particular attention was devoted to highlighting the value of these structures to artists and designers.

3

underneath it all

introduction

In the context of the designed environment, geometry is a determining constraint. Euclidean geometry is that branch of geometry used to gain knowledge relating to the structural elements, figures, shapes and forms in the two-dimensional plane and in three-dimensional space. It is believed that Euclid and other ancient Greek geometers were informed substantially by the wisdom of ancient Egyptian scribes and priests (Heath 1921). Knowledge of Euclidean geometry passed from ancient Greeks to ancient Romans, and subsequently to the Byzantines and various Islamic civilizations. By the Middle Ages in Europe, it appears that geometric knowledge was held in secret among select groups and handed down from generation to generation of master builders and artisans. Knowledge of various geometric constructions traceable to the time of Euclid can be of value in the twenty-first-century design context, and for this reason these and related constructions, used in the decorative arts, design and architecture, are explained and illustrated in this chapter.

various useful constructions

Although human experience is set largely in the reality of a three-dimensional world, visualization among designers, even in the early-twenty-first century, relies invariably on impressions made by pencil or pen on paper. Certain basic geometrical constructions, many relying on the use of only a pair of compasses and a straight edge, have been used by builders, architects, designers and artists for well over two millennia. Various constructions were discovered and applied as skeletal features to aid the decorating or making of objects or buildings, with the intention of achieving what appeared to be the most visually desirable and practically functional result. The choice of such constructions was probably developed through a process of trial, error, selection and refinement, drawing perhaps from measures found in nature. Several researchers have endeavoured to recover (and, in a few cases, apply) the geometrical rules governing various ancient constructions; examples include Hambidge (1926), Edwards (1932 and 1967), Ghyka (1946), Brunes (1967), Critchlow (1969), Kappraff (2000 and 2002) and Stewart (2009). In addition to simple figures such as circles, squares and regular polygons, certain other figures and constructions are of particular importance to designers, artists and architects. Amongst these are the so-called *vesica piscis,* Reuleaux polygons, the sacred-cut square, root rectangles, and the Brunes Star, all of which are discussed at length in this chapter. Related constructions are discussed in later chapters.

Lines articulate shapes. In a general sense it could be claimed that three basic shapes are capable, by themselves, or through combinations and variations, of producing all possible constructed figures. The three shapes are the circle, the square and the equilateral triangle. Each may express one or more orientations or directions: the circle, curved components; the square, straight, horizontal and vertical components; the equilateral triangle, straight, horizontal and diagonal components (Dondis 1973: 46).

As noted in Chapter 1, a line can be considered as the path between two points. When a line is rotated in the plane, with the first point fixed as a centre, an arc is traced by the second point; if rotated through an angle of 360 degrees, a circle is created with radius equal to the length of the line. In formal geometric terms, a circle is an area enclosed by a set of continuous points forming a line equidistant from a centre point. The boundary of a circle is known as a circumference. A radius is any straight line extending to the circumference from the centre point. A chord is a straight line that connects any two points on the circumference of a circle. A diameter of a circle is twice the length of the radius and is a straight line passing through the centre of the circle connecting two points on the circle's circumference. A diameter is thus the longest chord.

The circle is the easiest geometrical figure to draw with accuracy, and it is a vital component in the construction of many other geometrical figures. It has a multitude of uses in the visual arts, and it is associated with, amongst many other things, halos, rainbows, the marriage ring, the prayer wheel, rose windows in cathedrals, the Tibetan mandala and Neolithic stone circles. Without the circle, the discipline of geometry would be a lost discipline, with no real function or

meaning. Imagine a world without circles or any form of curved components. Circular forms of design (often referred to as rosettes) can be traced back to ancient times, their importance recognized in ancient Egypt, Babylon, Assyria, Greece and Rome. They are also found commonly in the modern urban environment (Figure 3.1a–c). It has

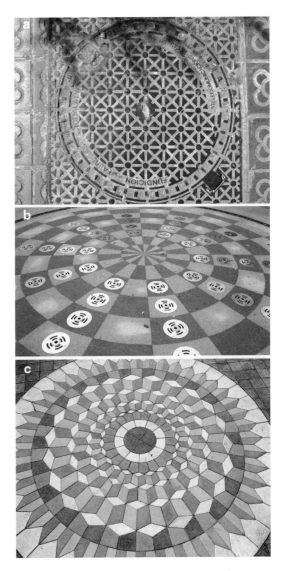

Figure 3.1a–c Manhole cover, Spain, 2008; Hanyang University, Seoul, 2010; Turkey, 2008.

been claimed that a large software corporation typically asked its short-listed candidates, at interview for posts with the company, why it is the case that most manhole covers are circular discs rather than square or some other regular geometrical form.

An important geometric configuration from the viewpoint of the visual arts is the *vesica piscis,* a construction made from two circles of the same radius, intersecting in such a way that the centre of each circle lies on the circumference of the other. The figure is relatively straightforward in terms of construction. First draw a straight line. With the point of a pair of compasses on the straight line, draw a circle. Using the same radius, draw a second circle, partly overlapping with the first (Figure 3.2). The area of overlap (enclosed within the emboldened arcs in the illustration) is the *vesica piscis.* Also known as a mandorla, because of its almond shape, the *vesica piscis* is of ancient origin, associated with Christian mysticism and found in use in Byzantine and Italian Renaissance times as a symbol of Christ (Calter 2000: 13). Four *vesicae* combined yield an eight-point star (Calter 2000: 12). The construction was known to ancient civilizations in India, Mesopotamia and West Africa. The mysticism associated

with the figure is explored briefly by Lawlor in his text *Sacred Geometry* (1982: 31–4).

The importance of the *vesica piscis* is that it can initiate other constructions. By way of example, Figure 3.3a shows the construction of an equilateral triangle and Figure 3.3b the construction of a regular hexagon. A third (equal) circle can be added (Figure 3.3c) and the resultant three-circle figure used as an alternative means of constructing a regular hexagon. The addition of two further (equal) circles yields a four-petal motif (Figure 3.3d). It is well known that six circles of a given size fit exactly around a seventh of the same size (Figure 3.3e). Six circles can be drawn around a central circle to produce a six-petal motif (Figure 3.3f). A further development is shown in Figure 3.3g. Connecting the centres of adjacent, close-packed or overlapping circles can produce various grids of value in tiling design (which is dealt with in Chapter 4).

Another class of construction worth noting is the Reuleaux triangle (and other Reuleaux polygons). If you examine a British twenty-pence or fifty-pence coin, you will see that the sides are not straight but curved. In fact, each side is an arc with the centre at the opposite corner. The Reuleaux triangle can be constructed in one of two ways: either by connecting the three centre points of three overlapping circles (shaded area of Figure 3.4) or by rounding each side of an equilateral triangle through the addition of an arc with a centre point at the opposite angle (Figure 3.5). This and similar figures (all from equal-sided polygons with odd numbers of sides) are named after a German engineer by the name of Franz Reuleaux (1829–1905).

In different cultures and during different time periods, the use of certain systems of proportion was allied closely to religiously prescribed

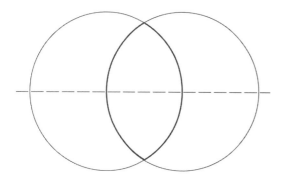

Figure 3.2 *Vesica piscis* (JSS).

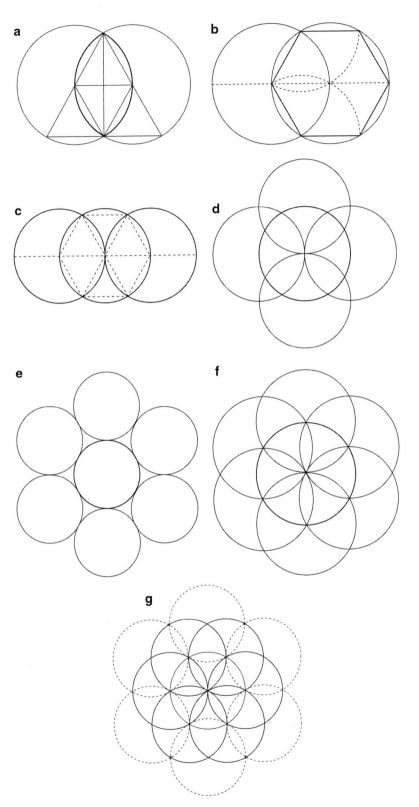

Figure 3.3a-g *Vesica piscis* and additional circles (JSS).

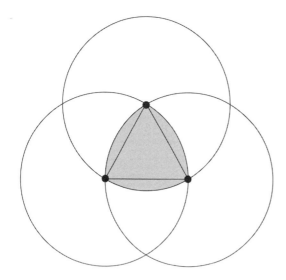

Figure 3.4 Reuleaux triangle 1 (JSS).

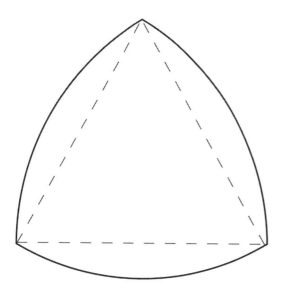

Figure 3.5 Reuleaux triangle 2 (JSS).

regulations; this was probably the case in ancient Egypt, Greece and Rome. In the context of architectural design, the challenge was to address both technical and aesthetic considerations as well as the necessity to ensure that resultant buildings were sympathetic to human proportion.

Marcus Vitruvius Pollio (ca. 70–25 BCE), commonly referred to as Vitruvius, was a Roman architect and engineer in the service of Emperor Augustus. Vitruvius produced a treatise (consisting of ten books) concerned with architecture and associated topics, much of which may well have been taken from earlier lost texts by Greek thinkers. Vitruvius's books contain the earliest surviving reference to the classical orders of architecture. Throughout much of his treatise, he stressed that knowledge of geometry and geometric construction was of great importance to the architect. Analysis of buildings at Pompeii and Herculaneum suggests that the design of many ancient Roman houses was based on systems of proportion associated with the square. One particular system of proportion is based on what is known as the sacred-cut square. Making reference to the work of the Watts (Watts and Watts 1986), Kappraff (2002: 28) described the use of this particular system of proportion and showed how, for example, its use was evidenced in parts of the excavated site of the Roman port of Ostia. Within the site, a complex known as the Garden Houses was unearthed, and its structure appeared to conform closely to the sacred-cut square. This is constructed as follows. First draw a square with two diagonals (lines connecting opposite angles). With a pair of compasses, centred on one of the corners of the square, open to half a diagonal (at intersection point with other diagonal) and draw an arc which cuts two perpendicular sides of the square. Repeat the process for the other three corners of the square, thus providing two intersections on each side. Draw two vertical lines and two horizontal lines which connect these intersections. A central square, known as the sacred cut, is thus formed (Figure 3.6). The sacred-cut construction can be continued inwards by repeating the procedure on the sacred-cut square. Kappraff (2002:

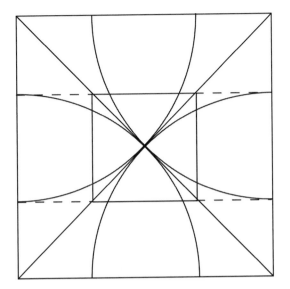

Figure 3.6 Constructing a sacred-cut square (JSS).

Figure 3.7 Diagonal on a square (JSS).

29, 32) showed that a double series of areas could be derived from a series of sacred-cut constructions and concluded that the system based on the sacred-cut construction was a successful system of proportionality and thus valuable in architectural and other construction.

static and dynamic rectangles

The principle of repetition of ratios, according to Kappraff (2002: 156), was well known in Italian Renaissance times and was of importance to the concept of dynamic symmetry and dynamic rectangles explained by Hambidge (1926). Hambidge's work was of great importance to the development of various concepts highlighted here. In *The Elements of Dynamic Symmetry* he introduced various categories of rectangular construction, based largely on a square and its diagonal, that is, a straight line from one interior angle to the opposite interior angle (Figure 3.7), and a square and its diagonal to its half, that is, a diagonal from one interior angle to the halfway point on one of the opposite

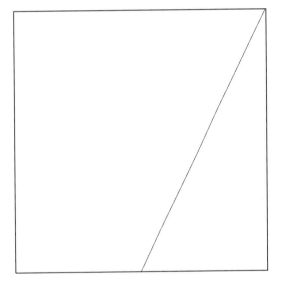

Figure 3.8 A half diagonal on a square (JSS).

sides (Figure 3.8). The first category is known as root rectangles and the second as the rectangle of the whirling squares; both were referred to as dynamic rectangles, which were differentiated from static rectangles. Static rectangles by comparison are formed simply by dividing a rectangle into even multiple parts, such as a square and a half, three quarters, one quarter, one third, two-thirds and so on. Dynamic rectangles appear to have been of great practical value and offer great potential to twenty-first-century practitioners across the full spectrum of design disciplines. With this in mind, further elaboration and explanation of their construction is presented in the following section.

root rectangles

The square and its diagonal are the basis for generating a series of rectangles which, according to Hambidge, were used by the ancient Greeks and Egyptians in the design of buildings, monuments and sculptures. These are known as root rectangles and are constructed as follows. In Figure 3.9, ABCD is a square with sides of unit length one. Since the diagonal of a square (with side equal to unity) equals the square root of two (or √2), the diagonal DB equals √2. Using D as centre and DB as radius, an arc can be drawn as shown (BF). The line DF equals DB or the square root of two (√2 = 1.4142). So this rectangle AEFD is known as a root-two rectangle. Successive rectangles can be constructed as shown using the lengthening diagonal as a radius, to generate progressively

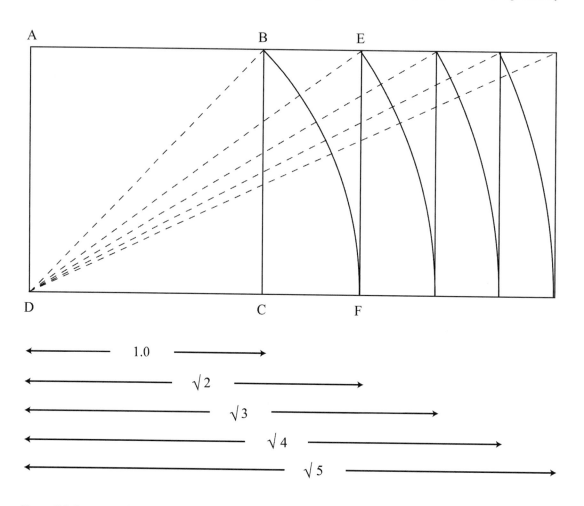

Figure 3.9 A square and root rectangles (JSS).

root-three ($\sqrt{3}$ = 1.732), root-four ($\sqrt{4}$ = 2) and root-five ($\sqrt{5}$ = 2.236) rectangles. So in each case the longer side of the root rectangle will be the root value and the shorter side the value of one. The process can be carried on to produce progressively higher orders of root rectangles, but for practical purposes, according to Hambidge (1967: 19), the root-five rectangle is the highest order found in design. Interestingly, if a square is constructed on the longer side of any of these root rectangles, its area will be an even multiple of a square constructed on the shorter side. For example, a square created with line DF (Figure 3.9), would have twice the area of a smaller square created with line AD. The same applies to the successive rectangles.

Individual diagrammatic representations for each root rectangle up to root five (constructed using the technique outlined previously) are presented in Figures 3.10–3.13. Division of rectangles in a particular way can also produce their reciprocals, identified by Hambidge, and seemingly used by the ancient Greeks. Hambidge commented that 'a reciprocal of a rectangle is a figure similar in shape to the major rectangle but smaller in size' (1967: 30). It can be seen from the illustrations that each root rectangle can be further subdivided to yield smaller root rectangles of the same order, and

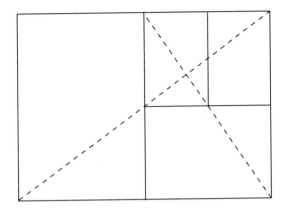

Figure 3.10 Root-two rectangle representation and subdivisions (JSS).

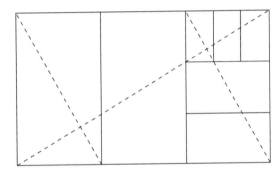

Figure 3.11 Root-three rectangle representation and subdivisions (JSS).

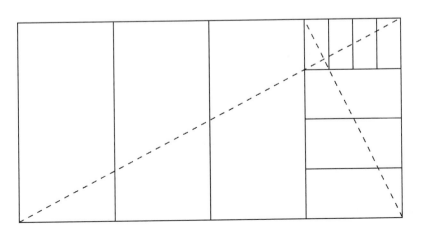

Figure 3.12 Root-four rectangle representation and subdivisions (JSS).

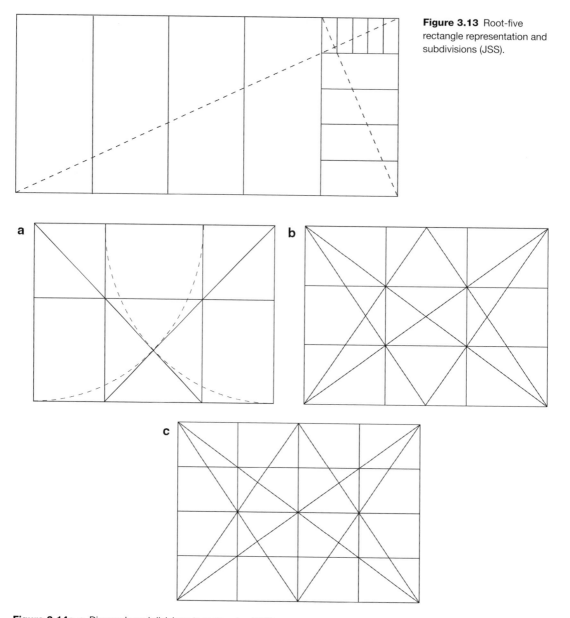

Figure 3.13 Root-five rectangle representation and subdivisions (JSS).

Figure 3.14a–c Diagonals and divisions to rectangles (JSS).

each of these subdivisions can be further subdivided to provide more of the same. These smaller rectangles are the reciprocals of the parent rectangle. All the root rectangles mentioned can yield interesting proportions and subdivisions, mainly through placing of diagonals and further divisions, as shown by way of example in Figure 3.14 a–c.

Hambidge introduced a particular rectangle derived from a construction initiated by drawing a diagonal to a square's half, as shown in Figure 3.15a.

He referred to the figure as the 'rectangle of the whirling squares', due to the series of successive squares that can be constructed within the figure itself (Figure 3.15b). Successive reciprocals create squares, and these squares are arranged in a spiral formation, 'whirling to infinity around a pole or eye' (Hambidge 1967: 31). Numbered areas 1, 2, 3, 4, 5, and so on, identify the successive (or 'whirling') squares created by the process. Possibly the most important feature of this rectangle (which will be discussed again in Chapter 6) and the various root rectangles is that each produces a range of further proportional figures.

Hambidge identified a substantial number of ratios, seemingly found in nature and used in ancient Greek constructions. Further to this he presented a series of rectangles constructed by reference to various ratios, reciprocals, half-ratios and half-reciprocals; the belief was that designers would find these of value as compositional aids.

Remarking on Hambidge's work, Ghyka stated:

The dynamic rectangles...can produce the most varied and satisfactory harmonic consonant, related by symmetry subdivisions and combinations, and this by the very simple process...of drawing inside the chosen rectangle a diagonal and the perpendicular to it from one of the two remaining vertices thus dividing the surface into a reciprocal rectangle...and then drawing any network of parallels and perpendiculars to sides and diagonals. This process produces automatically surfaces correlated by the characteristic proportion of the initial rectangle and also avoids (automatically again) the mixing of antagonistic themes. (1977: 126–7)

Ghyka thus observed that when proportions from the same parent sources are combined and used together within the one composition, they will have a tendency to work well visually. This is a crucial observation, and one worthy of note to twenty-first-century designers. By way of example a series of rectangles in the proportion of 1.6180:1 (the ratio of the whirling-squares rectangle) has been divided following closely systematic procedures proposed by Hambidge (1926), Ghyka (1946) and Elam (2001) (Figure 3.16).

It should be recognized that there is a tendency in the literature for researchers to believe that the one system of proportion which they have adopted as their focus is the one and only system of any value. This is plainly not the case, and

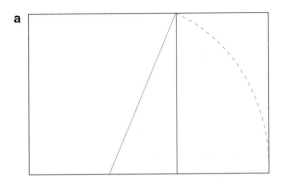
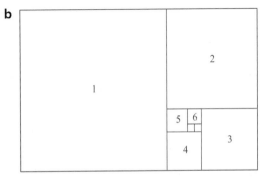

Figure 3.15a–b Diagonal to the half and the resultant whirling-squares rectangle (JSS).

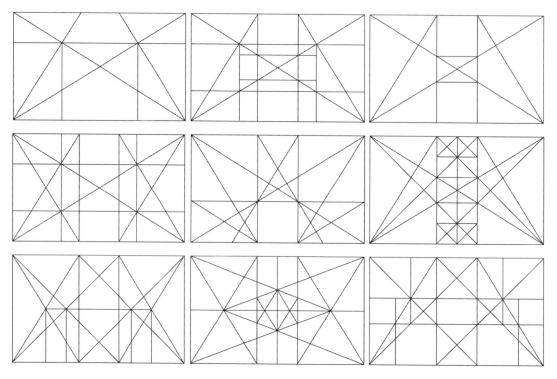

Figure 3.16 Various divisions of rectangles of the whirling squares (JSS).

Hambidge's system, for example, though clearly of immense potential value to the design analyst as well as the design practitioner, does not apply in all cases. Other research has shown the value of the circle or the square as the geometric initiators in the design process. Probably the most intriguing is the observations and commentary associated with a construction which has become known as the Brunes Star; this particular construction is the focus of the section that follows.

the Brunes Star

There has been debate over the historical use by designers, architects, builders and artists of certain geometric constructions. A construction which has become known as the Brunes Star is

one such construction. Named after Tons Brunes (1967), a Danish engineer, the construction is an eight-point star, occasionally used for decorative purposes (Figure 3.17) but more commonly as an underlying structural device. The Brunes Star is created through the subdivision of a square into four equal component squares and the addition of various half diagonals (Figure 3.18). Further full diagonals can be added also.

It can be seen that when lines are drawn through intersection points within the construction, various equal segments result (e.g. on either side of the mid-point lines connecting opposite sides) (Figure 3.19). Based on this observation, it has been argued that the construction was used in the absence of a standard system of measurement (e.g. Kappraff 2000). Another interesting aspect of the construction is that it can act as a basis

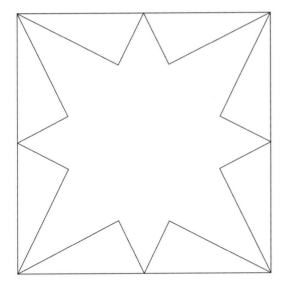

Figure 3.17 The Brunes Star (JSS).

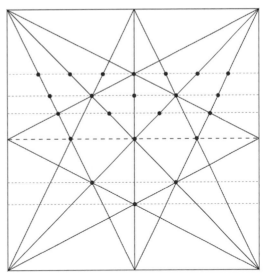

Figure 3.19 The Brunes Star with intersection points identified (JSS).

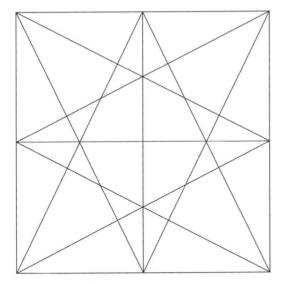

Figure 3.18 The Brunes Star divisions (JSS).

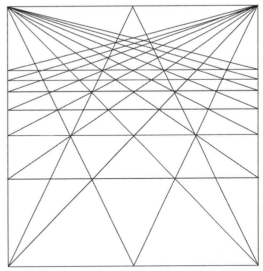

Figure 3.20 The Brunes Star and linear perspective (JSS).

for the construction of lines of linear perspective (Figure 3.20). The Brunes Star is relatively easy to construct in the present day but relies on using a perfect square, a construction which, it has been argued (by Kappraff and others), may not have been a straightforward matter in ancient times. In

discussions of the nature of the Brunes Star, Kappraff (2000) makes the interesting conjecture that, in the absence of knowledge of exactly how to construct a square, a series of 5-4-3 triangles (specifically the four larger 5-4-3 triangles ultimately found in the Brunes Star) was used with the intention of

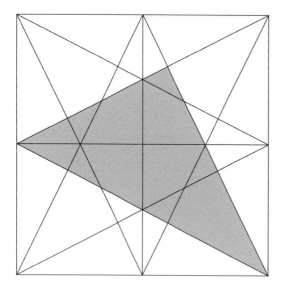

Figure 3.21 The Brunes Star with one of the larger 5-4-3 triangles identified (JSS).

simple rectangular components (e.g. the length and breadth of a poster, a book cover, a floor plan, a carpet's dimensions). Greater potential can however be realized.

When consideration was given to the Brunes Star in the previous section, it was seen that the construction relied on the addition of diagonals to a square, the subdivision into four equal component squares and the addition of various further diagonals. It is proposed here that similar construction lines be applied to rectangles constructed in the following side ratios: 1:1.4142 (the ratio of the root-two rectangle); 1:1.732 (the ratio of the root-three rectangle); 1:2 (the ratio of the root-four rectangle); 1:2.236 (the ratio of the root-five rectangle). So in each case, two

constructing a square, and the triangles were not therefore a by-product of the star's construction (Figure 3.21). It is accepted that knowledge of the construction of a square may not have been widespread in ancient times, but in the ascendancy of geometrical discovery in antiquity it is difficult to believe that using four equal-sized 5-4-3 triangles assembled in a very specific way to create a square predates knowledge of the construction of a square using a pair of compasses and straight edge. Further to this, common sense suggests that the various constituent 5-4-3 triangles of the Brunes Star were by-products of the construction rather than its building blocks.

compositional rectangles

The root rectangles offer immense potential as compositional aids to the creative practitioner, especially as important indicators for length-and-width proportion within a design. They may, for example, be used to determine the proportion of

Figure 3.22 Root-two rectangle with Brunes-Star-type construction (AH).

Figure 3.23 Root-three rectangle with Brunes-Star-type construction (AH).

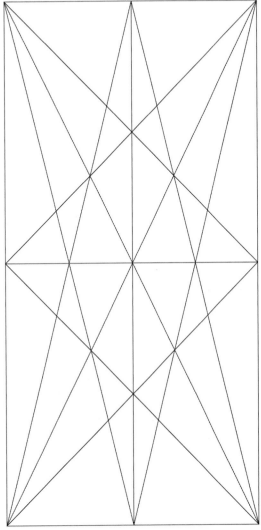

Figure 3.24 Root-four rectangle with Brunes-Star-type construction (AH).

corner-to-corner diagonals are drawn, and the centre of the rectangle is thus identified. Each angle at the centre is bisected and lines are drawn to connect the mid-points of opposite sides thus dividing the rectangle into four equal parts. Next, the eight half diagonals (two diagonals from each angle to the mid-points of the two opposite sides) are drawn. Each rectangle has thus been given a Brunes-Star-type division, though not based on a square. These divided rectangles can be referred to as compositional rectangles, in that they are intended for use in assisting with the placement of elements within any visual composition.

The intersection points (which we can call key aesthetic points or KAPs), where diagonals and other internal lines cross over each other, are ideal markers for the positioning of key elements, features or components of a design, structure or composition (Figures 3.22–3.25). Potential is thus offered

Figure 3.25 Root-five rectangle with Brunes-Star-type construction (AH).

communications, posters, paintings or sculptures); and as a basis on which to organize components of other, more substantial visual statements or plans, including building and landscape designs.

grids from dynamic rectangles

The value of grids in design organization was highlighted previously. A series of original grids for use by artists and designers is here presented, each with cells of equal size and based on one of the root rectangles (referred to by Hambidge as dynamic rectangles) used in the previous section. As noted in that section, these have the following ratios: 1:1.4142 (the ratio of the root-two rectangle); 1:1.732 (the ratio of the root-three rectangle); 1:2 (the ratio of the root-four rectangle); 1:2.236 (the ratio of the root-five rectangle). Each is presented here (Figures 3.26–3.29). These grids are of value in circumstances where one or more elements in a design undergo repetition. Like all regular grids (i.e. grids with equal-sized and equal-shaped cells), these root rectangles offer a guideline or framework for the disposition of elements, in order to achieve balance and unity. Further possibilities can unfold. While these grids are of great potential when used on their own, the reader is encouraged to select compatible scales (with cell walls of equal, twice or three times the length) and to combine the relevant root-rectangle cells with cells from other grid types, either on a like-with-like or on a mix-and-match basis.

as a compositional aid to assist with the construction of motifs for repetition and with the disposition of components within non-repeating visual compositions (e.g. Web-site pages, advertising

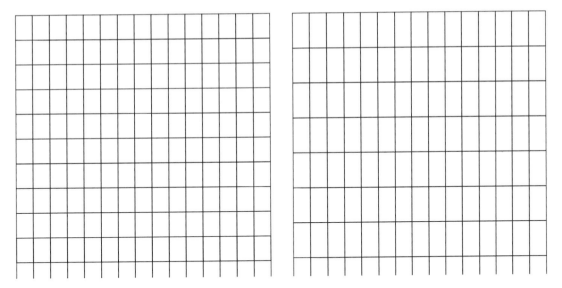

Figure 3.26 Grid with root-two unit cell (JSS).

Figure 3.28 Grid with root-four unit cell (JSS).

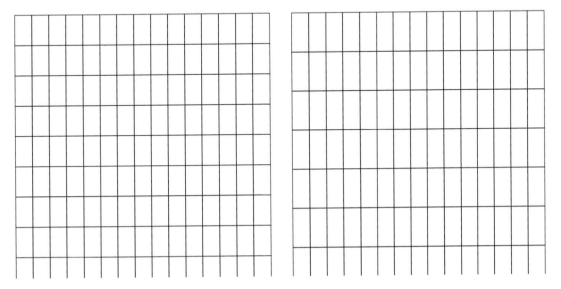

Figure 3.27 Grid with root-three unit cell (JSS).

Figure 3.29 Grid with root-five unit cell (JSS).

CHAPTER SUMMARY

This chapter has considered briefly a small number of geometrical constructions based on circles and squares, and has introduced root rectangles and the so-called Brunes Star. The work of Hambidge (1926) is a particular focus. A series of compositional rectangles (based on root rectangles, referred to by Hambidge as dynamic rectangles) and a system of related grids are proposed. These, together with similar proposals presented in other chapters, will form the basis of a designer's framework brought together in the concluding chapter.

Probably the most important realization that should come from this chapter is that, when components within a visual composition, design or construction are taken from the same geometrical source, they will stand a greater chance of being compatible and fitting aesthetically. Parts of a design can thus be made geometrically sympathetic to the whole of the design. In other words, component parts should show consistent proportionality with the whole, a concept referred to as geometric complementarity.

Systematic subdivisions of the various root rectangles can thus create structural frameworks of value to practitioners working in any branch of the visual arts, design and architecture.

4

tiling the plane without gap or overlap

introduction

The term *tiling* is used to refer to a special category of two-dimensional design created simply by dividing the plane (e.g. a sheet of paper) into a network of shapes. Unusually, the term tiling is derived from the material or carrier of the design itself, as most tilings are realized in physical form through assemblies of flat shapes (made typically from ceramic material, stone, clay or glass) known as tiles. Tiling assemblies may be periodic, where a constituent tile or a specific assembly of tiles exhibits regular repetition across the plane, or aperiodic, where regular repetition is not a characteristic feature. In both cases assembly is generally on a flat surface, and the constituent tiles cover the given area without gap or overlap (i.e. they tessellate).

Tilings may use one or more varieties of regular polygon (with equal interior angles and equal-length sides in each case) or may be assembled from a series of irregular polygons. In the context of this book the focus is on tilings rather than mosaics, though in terms of structural characteristics there can be substantial overlap between the two. Tilings are constructed on a strict geometric framework. They usually exhibit regular repetition (though, as noted, occasionally they do not) and are composed of quantities of a relatively small number of different shapes, assembled systematically. The resultant tiling design can be imagined stretching across the plane (invariably to infinity, so beyond any real-life measurable setting). Mosaics, on the other hand, are largely figurative visual statements, complete within clearly defined margins or boundaries and using as many colours and shapes as necessary in order to render the desired composition.

Tessellations can occur naturally (e.g. a honeycomb) or may be manufactured (e.g. stained glass, a ceramic floor, a textile or wall tiling). In the manufacturing context, prior to tilings being realized in physical form, a design sketch or plan needs to be produced, thus determining in advance the exact disposition of tiling elements, colours and dimensions. The first stage of this preliminary process involves partitioning the plane (e.g. a sheet of paper or a rectangle presented on a computer screen) with intersecting lines, typically drawn using a pencil, a pair of compasses and a straight edge or, alternatively, a computer mouse or other digital drawing implement.

Kepler, the renowned scientist and astronomer, carried out an early investigation into the mathematical aspects of tilings. In his publication *Harmonices Mundi Libri V* (1619) he presented a tiling which combined pentagons, pentagrams, decagons and fused decagon pairs. In the late-twentieth century much of what was known about the mathematics of tilings was presented by Grünbaum and Shephard in their monumental treatise *Tilings and Patterns* (1987). The majority of art or design students may find much of the text in Grünbaum and Shephard's work incomprehensible due to the barrier created by unfamiliar mathematical symbols or terminology; however, this impressive publication offers an unparalleled source of large numbers of network and tiling illustrations and, as such, is of great value as a visual reference for both the artist and designer in the development of original ways of dividing the plane within any composition (periodic

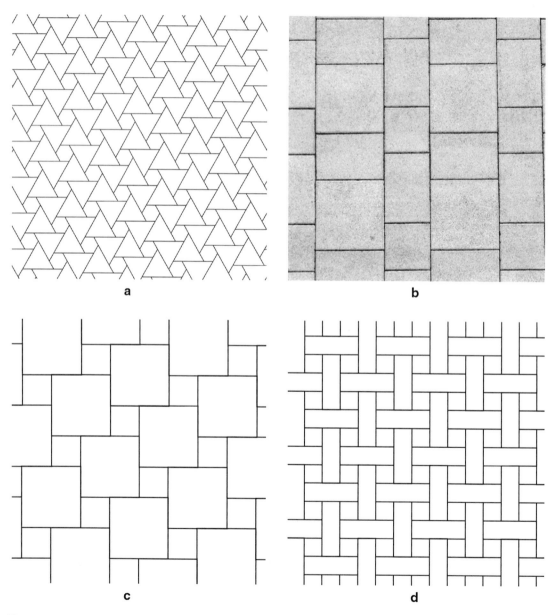

a

b

c

d

Figure 4.1a–d Non-edge-to-edge tilings (JSS).

or not). It is worth noting that a smaller paperback version of the treatise, consisting largely of key chapters with less obvious mathematical content, was published a few years later (Grünbaum and Shephard 1989).

As indicated previously, the focus of this chapter is on designs in which all components repeat in a systematic way and thus exhibit regularity, involving the use of one, two, three or more regular polygons. Attention is thus placed largely on periodic tilings, i.e. those which have a unit (consisting of one or more polygons) which is repeated in two independent directions across the plane in such a way as to ensure that gaps and overlaps are avoided. The basic forms of tilings are those created by combining regular polygons with equal-length sides, so they tessellate edge to edge, with the result that each polygon shares each side with one adjacent polygon only. Before turning attention to these, it is worth mentioning a particular subcategory of periodic tilings, those which do not tessellate edge to edge in the sense mentioned earlier and which may or may not use regular polygons with equal-length sides. Examples of these non-edge-to-edge assemblies are shown in Figure 4.1 a–d. It is however the edge-to-edge periodic tilings which receive the bulk of attention in this chapter. There are three distinct classes of these: regular tilings, semi-regular tilings and demi-regular tilings. As indicated previously, tilings which do not show regular repetition are known as aperiodic; within this category are Penrose-type tilings which cover the plane without gap or overlap and, although constructed following systematic rules, do not exhibit repetition. This variety of tilings is given some consideration in this chapter, and attention is given also to various types of tiling design associated with Islamic cultures, including star tilings (where the main components do not consist of regular polygons). Also of interest are so-called hyperbolic tilings, created within the boundaries of circles. A series of techniques for developing collections of original tiling designs based largely on grid structures is proposed also. Illustrative examples are provided from a selection of historical contexts.

regular and semi-regular tilings

As indicated, part of the concern of this chapter is with regular polygons which cover the plane, edge to edge, without gap or overlap. A regular tiling consists of copies of a single polygon of one size and shape. Only three types of regular polygon can, on their own, tessellate the two-dimensional plane, edge to edge, without gap or overlap: equilateral triangles (Figure 4.2a), squares (Figure 4.2b), and regular hexagons (Figure 4.2c), each consisting of partitions of the plane into equal-sized cells. What about tilings consisting of just regular pentagons or heptagons (Figure 4.3 and 4.4)? These and other higher-order regular polygons are unable to tessellate when used on their own. The explanation is as follows. In order to understand the restriction to only equilateral triangles, squares and regular hexagons, it is necessary to consider the interior angles of the constituent polygons at the point at which they meet in the plane. Polygons can only tessellate when the sum of the angles that meet at this point (known as the vertex) is equal to exactly 360 degrees. Six equilateral triangles (with each of the six having angles of 60 degrees), four squares (with each of the four having angles of 90 degrees) and three hexagons (with each of the three having angles of 120 degrees) fulfil this requirement (Figure 4.5). In each of the three cases, *vertices* (a term used in this book

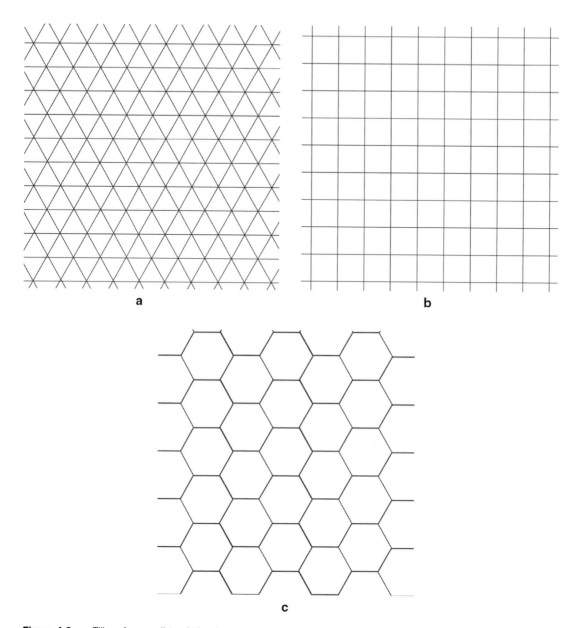

a

b

c

Figure 4.2a–c Tilings from equilateral triangles, squares and regular hexagons (JSS).

to denote points where angles of different polygons meet) are closed (with angles adding up to exactly 360 degrees) and are identical throughout the design. Pentagons (five sides with interior angles of 108 degrees), and heptagons (seven sides with interior angles of 128.6 degrees) cannot fulfil the 360-degree vertex requirement. Nor can octagons (eight sides with interior angles of 135 degrees), enneagons (nine sides with interior angles of 140 degrees), decagons (ten sides with interior angles of 144 degrees), hendecagons (eleven sides with interior angles of 147.3 degrees) and

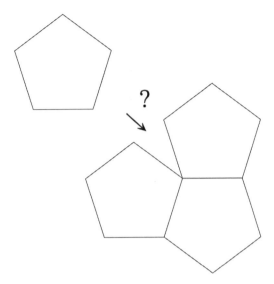

Figure 4.3 Pentagons do not fit (JSS).

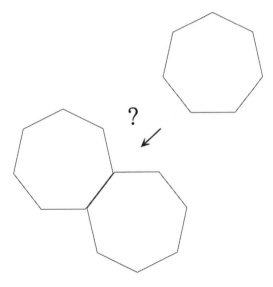

Figure 4.4 Heptagons do not fit (JSS).

dodecagons (twelve sides with interior angles of 150 degrees).

There are several forms of notation in use. These indicate the number of sides of each polygon and the number of polygons which meet at the vertex. In each of the three cases there is only one type of vertex. A tessellation of squares can be notated by (4, 4); that is, each square has four sides, and there are four squares at each vertex. The tessellation of regular hexagons can be notated by (6, 3) and that of equilateral triangles by (3, 6).

Tessellations of the plane using two or more different regular polygons are also possible. Confusingly, this next class of tilings (consisting of combinations of regular polygons) is known as the semi-regular tilings. There are eight possibilities which fulfil both the 360-degree-vertex rule and have only one vertex arrangement within the design. The eight vertex arrangements are as follows: four equilateral triangles and one hexagon (Figure 4.6a); three equilateral triangles and two squares (two possibilities, shown in Figures 4.6b and 4.6c); one equilateral triangle, two squares and one hexagon (Figure 4.6d); two equilateral triangles and two hexagons (Figure 4.6e); one equilateral triangle and two dodecagons (Figure 4.6f); one square, one hexagon and one dodecagon (Figure 4.6g); one square and two octagons (Figure 4.6h).

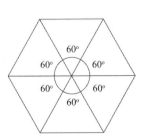

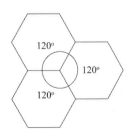

Figure 4.5 The three that fit (JSS).

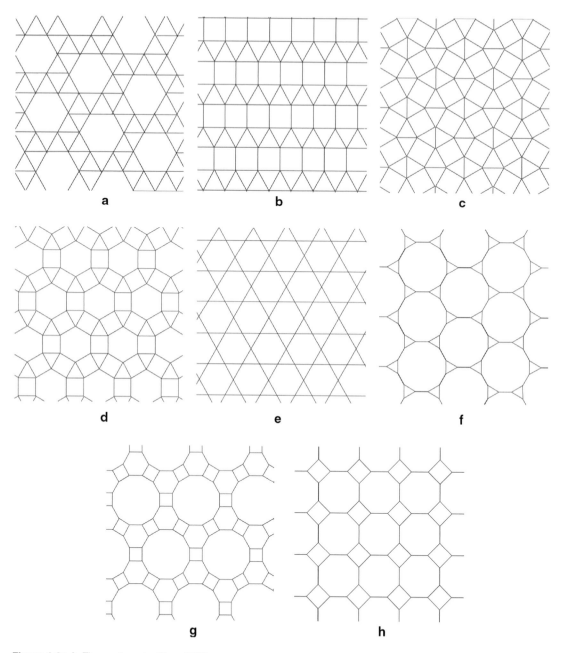

Figure 4.6a–h The semi-regular tilings (JSS).

demi-regular tilings

An agreed-upon definition of a demi-regular tiling cannot be found, although there seems to be a common perception that these are simply tilings with more than one vertex type. A few authors have maintained that there are fourteen possibilities (Critchlow 1969: 62–7; Ghyka 1977: 78–80; Williams 1979: 43). However, there appears to be disagreement on the exact structure of all

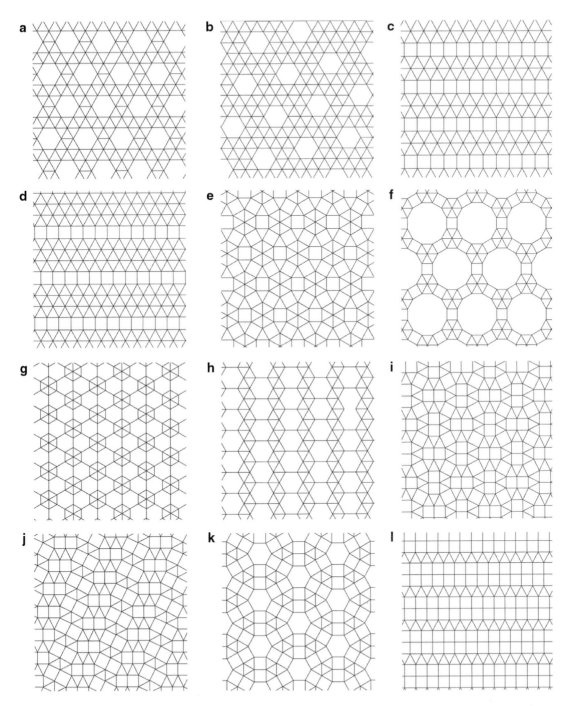

Figure 4.7a–t The demi-regular tilings (JSS) (continued overleaf).

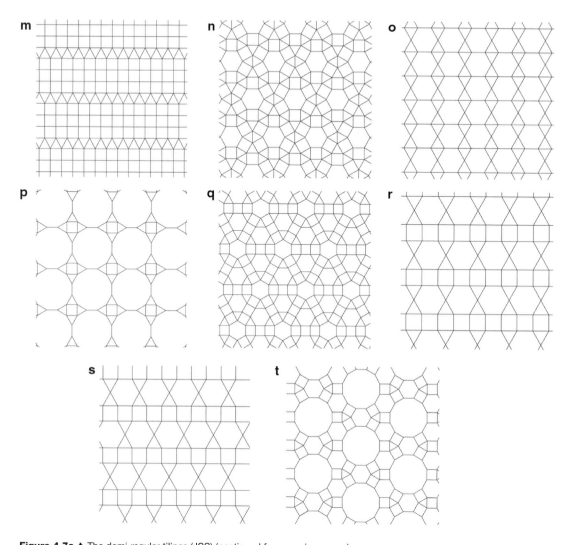

Figure 4.7a–t The demi-regular tilings (JSS) (continued from previous page).

fourteen. For a definitive answer, it is necessary to make reference to Grünbaum and Shephard, who use the term 2-uniform tessellations and present illustrations for a total of twenty tilings, each with two vertex types (1989: 65–7). Reproductions of these twenty tiling designs are presented in Figure 4.7a–t. Such structures are invaluable to designers in the quest to develop fresh repeating designs. It is particularly worth considering each tiling design as a grid on which to arrange other designs. Also some of the techniques identified in Chapter 2 for generating alternative but related structures should be considered.

There are numerous additional possibilities, and readers wishing to explore these and access a strong, academically focused treatment of the subject are referred to Grünbaum and Shephard (1987 and 1989). Readers who may wish simply to access a useful collection of tiling alternatives should refer to the paper by Chavey (1989). Many of the illustrations provided in this chapter were reproduced by making reference to these and other related publications from the substantial variety of mathematically oriented literature.

aperiodic tilings

The regular, semi-regular and demi-regular tilings examined in the previous sections are periodic (i.e. they exhibit regular repetition in two distinct directions across the plane) and are assembled edge to edge without gap or overlap. There is also a class of tilings which does not exhibit regular repetition but nevertheless covers the plane without gap or overlap (i.e. they are aperiodic tilings). A particular class of these is known as Penrose tilings, named after Roger Penrose, the eminent British mathematician. Penrose tilings are constructed using polygons known as kites (with interior angles of 72, 72, 72 and 144 degrees) and darts (with interior angles of 36, 72, 36 and 216 degrees). There are further possibilities requiring only two types of polygon (or prototiles as they are referred to), one using two types of rhombus (or rhomb) with equal-length sides but different angles: a thin rhomb with interior angles of 36, 144, 36 and 144 degrees (Figure 4.8a) and a thick rhomb with angles of 72, 108, 72 and 108 degrees (Figure 4.8b). Quantities of this latter set can be readily used to create a periodic or repeating tiling, so certain strict rules in their assembly are required for the completion of an aperiodic assembly (Figure 4.9).

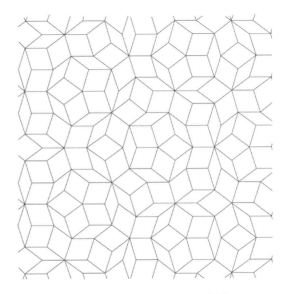

Figure 4.9 A Penrose-type tiling assembly (JSS).

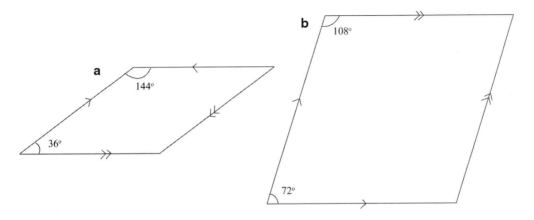

Figure 4.8a–b A thin rhomb; a thick rhomb (JSS).

Islamic tilings and their construction

Decoration consisting of tiles covering flat surfaces existed in ancient times in many cultural contexts, including ancient Rome and the Sassanian and Byzantine empires. Such decoration was appropriated, elaborated and developed to impressive heights of aesthetic sophistication in various Islamic cultures. The instruments used for design construction, in imitation of Greek mathematicians, were the pair of compasses and the straight edge, though further practical aids such as stencils and set squares (Sutton 2007: 4) were seemingly employed also. Underlying structural elements or frameworks included circles and regular, semi-regular and demi-regular grids identical to those formed by the regular, semi-regular and demi-regular tilings identified in the previous sections. Probably the most common of these were the regular tilings (equilateral triangles, squares and hexagons). Many designs were fully repeating tessellations and, when complete, tiled the plane with regularity and without gaps or overlaps. Some tiling designs were edge-to-edge and some were not. Other designs were non-repeating compositions created specifically within the confines of a rectangle or other regular or irregular shape. In many cases, other visual devices, including occasional figurative or floral motifs of various kinds, arabesques, calligraphy or interlacing effects, were incorporated, though in each case there was still adherence to a strict underlying geometric grid structure. It is this geometric grid or strict division of the plane which is a fundamental structural aspect of Islamic tiling design and construction, though it should be noted that evidence of the precise means and stages of construction of historical designs is limited.

Early examination and illustration of Islamic designs (including several tilings) were provided by Owen Jones (1856) in his renowned work *The Grammar of Ornament.* Probably the most notable and widely used early catalogue of Islamic tilings was produced some years later by Bourgoin (1879) under the title *Les éléments de l'art arabe: le trait des entrelacs;* in the late-twentieth century the work was made available under the title *Arabic Geometrical Pattern and Design* though without the original text (Bourgoin, 1973). Since Bourgoin's initial publication, there have been numerous publications in the subject area of Islamic tiling geometry. Not surprisingly the bulk of these has been produced by mathematicians, though in many cases there were attempts (with varying degrees of success) to make the findings understandable also to a non-mathematical readership. Among the most notable have been Hankin (1925), Critchlow (1976), El-Said and Parman (1976), Wade (1976), Grünbaum, Grünbaum and Shephard (1986), Lee (1987), Wilson (1988), Chorbachi (1989), Abas and Salman (1995), Necipoglu (1995), Sarhangi (2007), Castéra (1999), Kaplan (2000), Özdural (2000), Abas (2001), Tennant (2003), Kaplan and Salesin (2004), Field (2004), Sutton (2007) and Lu and Steinhardt (2007).

In his treatise on Islamic geometric patterns, Broug (2008) presented detailed instructions on the steps to be taken in constructing designs associated with a range of well-known Islamic buildings, monuments and designed objects. A useful CD included with the book gives the construction sequences for all the designs analyzed and redrawn within the book itself, together with substantial further information, templates and image gallery. This is one of the most informative publications from which to develop knowledge of straightforward stages in constructing Islamic-type tilings.

An early treatise dealing with the drawing of geometric designs in what was referred to as Saracenic art was produced by Hankin (1925). He recognized the importance of grids as a basis for construction historically, and in his quest to reconstruct various design types commented: 'It is first necessary to cover the surface to be decorated with a network consisting of polygons in contact' (1925: 4). In his descriptions of designing 'hexagonal' and 'octagonal' tilings, he recognized the usefulness of using a diamond-shaped grid and a square grid respectively as a structural framework for construction (1925: 4–6, 6–10). He commented further: 'The original construction lines are then deleted and the pattern remains without any visible clue to the method by which it was drawn' (1925: 4). It does indeed seem to be the case that the structural guidelines used in the creation of Islamic designs were removed during construction, though Hankin notes a rare occasion where he discovered scratches on plaster work, and found these to be 'parts of polygons, which, when completed, surrounded the star-shaped spaces of which the pattern was composed, and it turned out that these polygons were the actual construction lines on which the pattern was formed' (1925: 4).

On the one hand, there is sufficient evidence to confirm the use of grids of various kinds (e.g. a few instances can be detected in ancient documents such as the Topkapi Scroll, held in Topkapi Palace Museum, Istanbul). On the other hand, the precise stages and means by which designs were created using a given grid have been lost with the passage of time. Nevertheless, Hankin (1925) and various other scholars such as Critchlow (1976) have over the years given convincing accounts of the probable stages of geometric construction.

In general everyday use, the term *arabesque* refers to more-or-less continuous patterns, either all-over or as frieze designs, consisting of visual components derived from plant sources (ivy-type tendrils, leaves, stems and buds are typical features). Although floral in terms of theme, such designs have a clear geometric underpinning. In Hankin's case, however, the term *geometric arabesque* was applied to what many years later were referred to simply as Islamic star patterns (e.g. Kaplan and Salesin 2004). Abas and Salman (1995), in their treatise on symmetry in Islamic tiling designs, commented: 'The most striking characteristic of Islamic geometric patterns is the prominence of symmetric shapes which resemble stars and constellations' (1995: 4). It is these dominant component motifs which is the focus of attention here. They can be found commonly with five, six, eight, ten, twelve and sixteen points (or rays), while orders of seven and nine, though rare, can be noted occasionally. Higher orders, generally in multiples of eight to ninety-six, are not unknown.

The construction of star-type motifs involves the overlapping of regular polygons within a circle, or the drawing of lines connecting pre-determined points on a circle (Figure 4.10a–f). The stages of construction involved typically with an eight-point star are shown in Figure 4.11 and the fully repeating development in Figure 4.12 (note the emergence of the cross shape formed between four star shapes). An eight-point star as a repeating unit in frieze-design form is shown in Figure 4.13. It is important to realize that there are several construction methods which could conceivably be employed with the same outcome. Field (2004) gave a particularly accessible explanation of Islamic design and construction showed various readily understandable construction methods for

various star-type motifs, and also presented a series of drawings of designs gathered from a wide
range of historic sites and other sources. Following procedures suggested by Field's (2004) work,
varieties of twelve-point construction are shown
in Figure 4.14a–h. The Alhambra Palace complex in Granada (Spain) is renowned for its tiling
designs, and this source has acted as an inspiration to generations of designers and artists
(including M. C. Escher). An example of a well-
known Alhambra design is provided with outline
in Figure 4.15a–b.

Girih tilings are a particular type of decorative
tiling associated historically with Islamic architecture. They are produced in sets of five distinct
shapes, with sides on all five tiles the same length:
a regular pentagon, with five interior angles each
of 108 degrees; a rhombus with interior angles of
72, 108, 72 and 108 degrees; a so-called bow-
tie tile (a six-sided figure known as a non-convex
hexagon) with interior angles of 72, 72, 216, 72,
72 and 216 degrees; a regular decagon with ten
interior angles, each of 144 degrees; an irregular
convex (or elongated) hexagon with interior angles
of 72, 144, 144, 72, 144 and 144 degrees. It is
interesting to note that all angles are multiples of
36 degrees, allowing many alternative combinations around a 360-degree vertex. Substantial
variation in design is thus offered. It is of further
interest to note that all the angles represented in
the two types of Penrose tiling mentioned previously (illustrated in Figures 4.8a, 4.8b and 4.9) are
represented also in the girih five-tile set. In fact, Lu
and Steinhardt (2007) recognized various similarities between girih tilings and Penrose tilings. This
is particularly intriguing since the use of girih tiles
can be traced back to various fifteenth-century
buildings such as the Darb-I Imam shrine in Is-
fahan (Iran), built in 1453 and selected by Lu as
an example to aid comparison between girih and

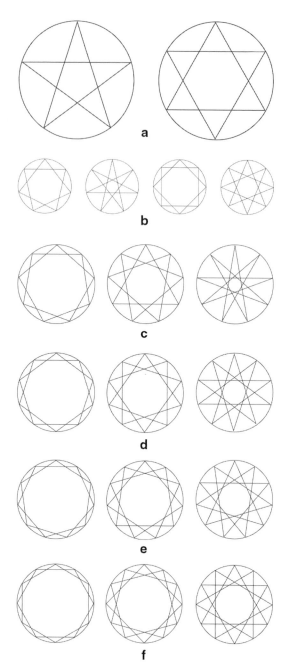

Figure 4.10a–f Star-type motifs (JSS).

Penrose-type tilings. The findings of Lu and Steinhardt (2007) were supported further by reference
to designs depicted in a fifteenth-century Persian
scroll known as the Topkapi Scroll (mentioned

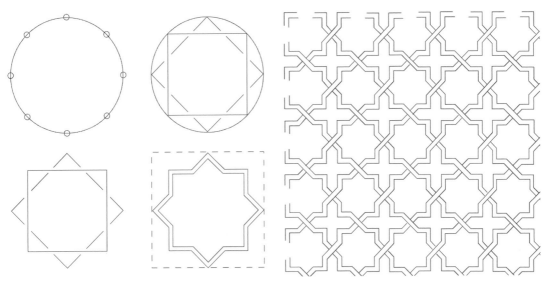

Figure 4.11 Construction of typical eight-point star (JSS).

Figure 4.12 Eight-point star in repeating form (JSS).

Figure 4.13 Eight-point star as a repeating frieze/border unit, Topkapi Palace Museum, Istanbul, 2010.

earlier in this section). A further characteristic of girih tilings which is worth noting is that each side of each tile is intersected at its mid-point by decorated bands at fixed angles; so when one tile is laid beside another, a band crosses from one to the other which gives the impression in the finished tiling of an interlacement effect weaving through the whole design. Such dramatic all-over effects, brought about by combinations of a relatively small number of constituent units, is a typical feature of Islamic tilings and is also a characteristic of modular arrangements (discussed in Chapter 9).

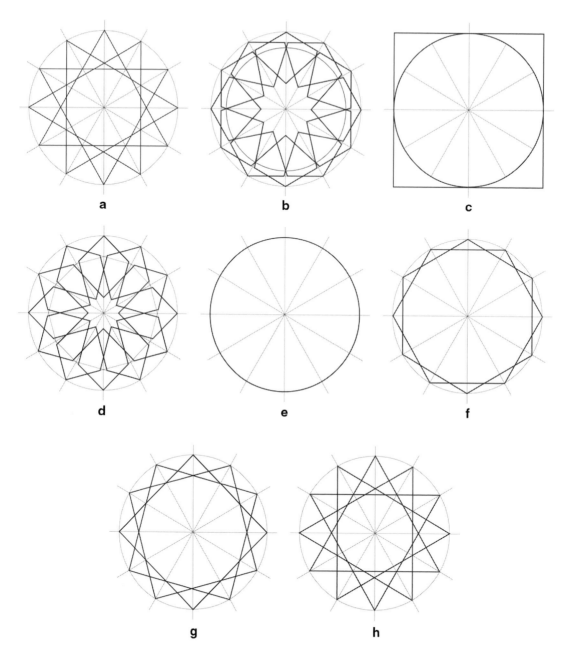

Figure 4.14a–h Various twelve-point star constructions (JSS).

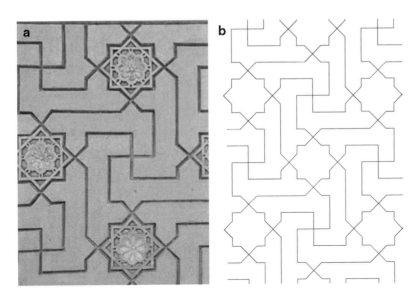

Figure 4.15a–b Alhambra Palace complex, Granada, Spain, photograph by Kholoud Batarfi (JSS).

tiling on a curved surface

The geometry of curved surfaces is known as hyperbolic geometry. Several researchers have considered tiling such surfaces. Hyperbolic tessellations cover hyperbolic space without gap or overlap. Regular polygons in hyperbolic geometry have angles smaller than they do in the plane, so the rules governing the plane tilings (addressed earlier) need to change. Tilings of curved surfaces are best explained and imagined using what is known as the Poincaré-disc model, a circular model which distorts distance in order to give the impression of a curved surface. So a motif or tile of a given size is depicted largest towards the centre of the circular disc (Figure 4.16). The artist M. C. Escher used concepts associated with hyperbolic geometry in the production of his series of four Circle Limit designs. Initially, Escher was inspired by an article sent to him by the Canadian mathematician Coxeter. The article included an illustration of a triangular design which used the Poincaré-disc model, of the type shown in

Figure 4.17. The underlying geometry of such designs has been described in an accessible way by Kappraff (1991: 420–1).

developing original tiling designs

A series of systematic means can be proposed which allows designers to generate frameworks from which to create tiling designs. Various regular, semi-regular and demi-regular tilings can act as a basis to create these frameworks. In most cases the tilings act as a grid, and the constituent cells are adjusted, moved, overlapped, expanded, combined or cut. A series of largely systematic procedures is presented next, some built from suggestions made by Willson (1983: 9–14), Wong (1972) or Kappraff (1991).

Duals and systematic overlaps

The dual of a tiling is obtained by joining the centres of each tile with the centres of all immediately

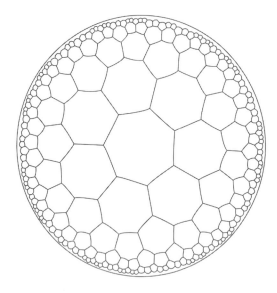

Figure 4.16 Tiling on the curve (JSS).

Figure 4.17 Poincaré-disc-model-type construction (JSS).

potential, however, can be realized by overlapping the dual tiling with the parent source and either keeping vertices at cell centres or manipulating them so that each is off centre (Figure 4.18a–d).

Removal and regroup

This involves systematic removal of certain tiles in one of the classified tilings in order to produce various regular all-over effects (Figure 4.19 a–d). These results can be enhanced further with the systematic addition of various simple textural effects or through systematic colour change. In all cases, it is important to be systematic and to ensure that repetition of components is retained.

Grid on grid

Similar to the technique proposed in 'Duals and systematic overlaps', this involves superimposing one classified tiling (or grid), introduced earlier in this chapter, on top of itself or another classified tiling (or grid) (Figure 4.20 a–b). Again, being systematic is crucial in order to ensure a result which is satisfactory visually. As with other procedures, the results can be enhanced through the systematic addition of texture and colour.

Combination alternatives

This technique involves combining two or more areas in a tiling by the systematic removal of one or more sides of selected tiles or common sides of cells in a grid (Figure 4.21). Tiles are thus made larger (e.g. two equilateral triangles can form a diamond shape, two diamonds a chevron and three diamonds a honeycomb).

adjacent tiles. The duals of the regular tilings do not offer much immediate potential with the square self-dualling, and with a tessellation of equilateral triangles being the dual of the hexagon-based regular tiling and vice versa. The

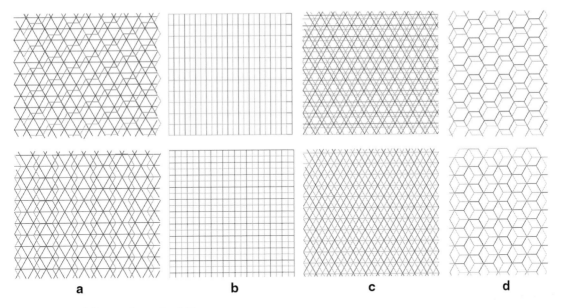

Figure 4.18a–d Systematic overlap (MV).

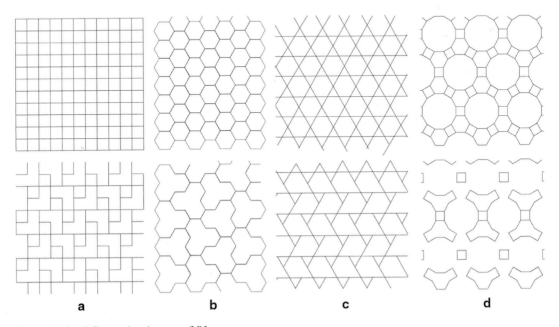

Figure 4.19a–d Removal and regroup (MV).

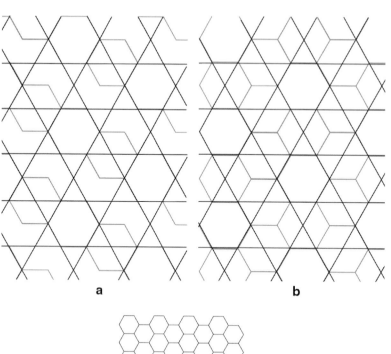

Figure 4.20a–b Grid on grid (MV).

a   b

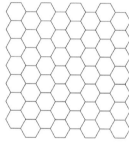

Figure 4.21 Combination alternative (JSS).

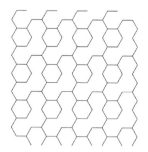

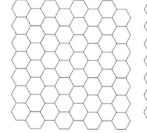 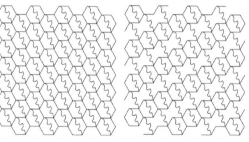

Figure 4.22 Systematic further division (MV).

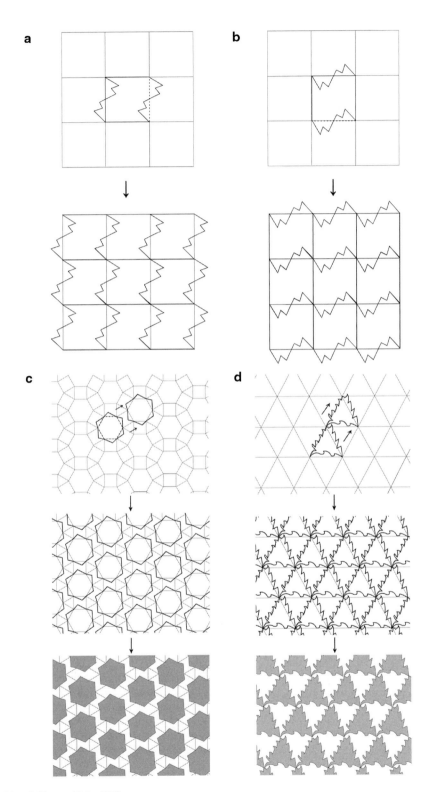

Figure 4.23a–d Give and take (JSS).

Figure 4.24 A collection of surface-pattern designs (Muneera Al Mohannadi).

Systematic further division

This involves adding or taking away lines to constituent cells within a classified tiling (Figure 4.22). Again, it is important to be systematic in order to ensure a result which is acceptable visually.

Give and take

This technique involves adding to sides and taking away the same shape from opposite sides, thus ensuring a continuous tessellation with precise registration between all tiles (Figure 4.23 a–d).

Cut, colour, rearrange and repeat

With this technique, used for producing a repeating tiling, a regular polygon is divided into three or more equal or unequal parts, and these parts are copied, coloured and reassembled in a repeating format. An example of a student response to an assignment, based on this and related procedures, is shown in Figure 4.24. The details of the relevant assignment are given in Appendix 1: Modular Tiling.

CHAPTER SUMMARY

A tiling is best considered as a network of shapes which cover the plane without gap or overlap. Tilings may be periodic, and show regular repetition of a component part, or aperiodic, where coverage of the plane without gap or overlap is a feature but regular repetition is not. Various categories of tilings were classified, including the regular, semi-regular and demi-regular tilings, differentiated in terms of the number of types of equal-shaped and equal-sized polygons used in each and by the number of equal vertices (the junction where the angles of component tiles meet). Attention was turned briefly to Islamic tilings and their construction, and also to ways of representing tilings on curved surfaces. A series of systematic procedures was proposed, aimed at stimulating the creation of original tiling constructions among modern-day designers. Examples of student work inspired by Islamic tiling designs are presented in Figures 4.25–4.31.

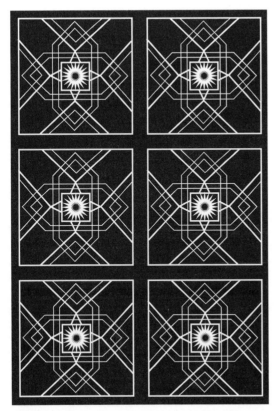

Figure 4.25 Design (1) inspired by Islamic tiling designs (Nazeefa Ahmed).

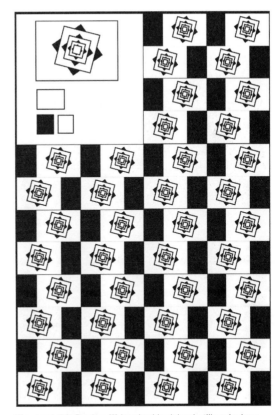

Figure 4.26 Design (2) inspired by Islamic tiling designs (Nazeefa Ahmed).

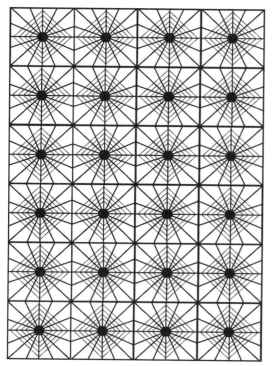

Figure 4.27 Design (3) inspired by Islamic tiling designs (Nazeefa Ahmed).

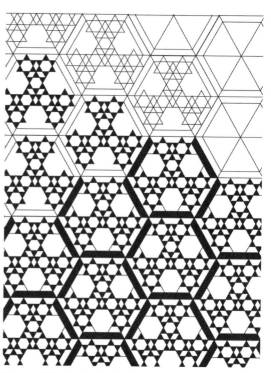

Figure 4.28 Design inspired by Islamic tiling designs (Elizabeth Holland).

Figure 4.29 Design inspired by Islamic tiling designs (Olivia Judge).

Figure 4.30 Design inspired by Islamic tiling designs (Alice Hargreaves).

Figure 4.31 Design inspired by Islamic tiling designs (Reerang Song).

5

symmetry, patterns and fractals

introduction

Both natural and man-made environments are dominated by order. Order is apparent in the thought processes of most humans and in the behaviour of most living creatures. The perception of order is central to our understanding of the world around us. There is order in language, in the seasons, in the landscape, in the growth of all plants and animals, in music, in poetry, in mathematics, in all branches of science and engineering, in dance, and in the visual arts, design and architecture. Order is a state of non-randomness, with everything in a logical, systematic sequence and seemingly adhering to a plan. Order is a characteristic of symmetry, the principal concern of this chapter.

Symmetry is a feature governing structure in a vast range of forms at the nano-, micro- and macro-levels in the natural and constructed worlds. While definitions vary to some degree, balance of physical form (including the forces underpinning and determining the shape of that form) seems to be an important inherent characteristic. In everyday use, the term *symmetry* is applied to a form which exhibits two equal parts, each a reflection of the other (as if in an imaginary mirror). This is bilateral symmetry, a characteristic of the majority of designed objects and constructions. However, the concept of symmetry can be extended beyond this meaning to include figures or objects consisting of more than two parts of identical size, shape and content. Examples of published work highlighting the all-encompassing nature of symmetry include Weyl (1952), Rosen (1989), Senechal (1989), Abas and Salman (1995: 32–5), Hahn (1998) and Wade (2006). Probably most enlightening of all are the two compendia edited by Hargittai (1986, 1989), consisting of over one hundred papers from the sciences, arts and humanities, each paper focused on a particular aspect of symmetry. Washburn and Crowe (1988) produced an impressive treatise dealing with the theory and practice of pattern analysis and explaining how symmetry concepts could be used in the analysis of designs from different cultural settings and historical periods. This has proved to be an important reference for anthropologists, archaeologists and art historians. A later publication by the same authors explored how cultures have used pattern symmetry to encode meaning (Washburn and Crowe 2004). Schattschneider produced the definitive study of the work of M. C. Escher, focusing in particular on the symmetry aspects of the artist's work, while at the same time providing explanations which were understandable to readers without specialist mathematical knowledge (Schattschneider 2004). Scivier and Hann explained the principles of layer symmetry and focused on their relevance to the classification of woven textiles (2000a, 2000b). A readily understandable explanation of symmetry in the realms of three-dimensional crystallography was provided by Hammond (1997). The emphasis in most of the available studies of symmetry in pattern is on analysis rather than synthesis and construction.

As is the case with the definitions of both point and line, symmetry can be considered to have internal structural meaning and significance on the one hand as well as external function and application on the other. Thus, symmetry can be understood as a product of a transitional process involving the interplay often of identical components (e.g. figures, surfaces, forms, bands of energy or forces of one kind or another) continuously mapping on to one another, almost as if the process was invisible to the naked eye, which sees only the external manifestation, a quiet status quo. In the scientific or mathematical context, the internal or underlying geometric meaning appears to be of significance, whereas, when considering symmetry and its perception in the context of the decorative arts and design, it is the external function or application which is of importance. In this latter case, symmetry is understood as a stable state arrived at through combinations of certain actions or forces known as symmetry operations (described in the following paragraph). Symmetry is thus a unifying principle with applicability across an impressive domain. Invariably, symmetry involves regularity, equality, order and repetition, and is an important aspect of structure and form in art, design and architecture. Meanwhile, the opposite state, known as asymmetry, is a characteristic of irregularity and disorder (though not all asymmetrical structures are necessarily irregular and disordered). Symmetry is an organizing principle that imposes constraints on physical construction. In the area of the decorative arts, symmetry concepts have proved to be of particular value

in characterizing two-dimensional figures, motifs and regularly repeating designs. Although of relevance in the classification of three-dimensional phenomena such as crystals, the concepts are most readily understood in the context of two-dimensional repeating designs and their component parts. For this reason, attention is focused in this chapter on explaining the applicability of symmetry in the two-dimensional context and, in particular, on how these forces known as symmetry operations can ensure that the component units of a regularly repeating design fit together and repeat with strict regularity.

Symmetry in design, especially regularly repeating designs, is considered invariably in terms of the design's underlying geometry and the various symmetry operations mentioned earlier. There are four of these relevant to two-dimensional design: rotation, reflection, translation and glide reflection (all shown schematically in Figure 5.1). Each is explained in turn in this chapter. Reference is made also to three design types: motifs, which are either free-standing designs or the recurrent components or building blocks of regularly repeating designs; frieze patterns (also known as strip or border patterns), which show regular repetition of a motif or motifs in one direction across the plane; all-over patterns (also known as plane or wallpaper patterns), which show repetition of a motif or motifs in two directions across the plane. All three design types can be classified in terms of their constituent symmetry characteristics. This chapter describes and illustrates the four symmetry operations and how two-dimensional designs may be classified by reference to symmetry characteristics. The related topics of counter-change designs (which rely on shade or colour change associated with a particular symmetry operation) and fractals (phenomena which exhibit repetition

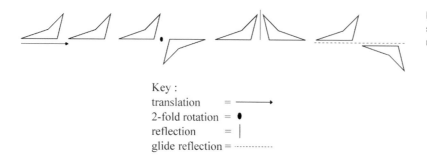

Figure 5.1 The four symmetry operations and relevant key (JSS and MV).

Key :
translation   = ⟶
2-fold rotation = ●
reflection   = |
glide reflection = ⋯⋯⋯

of a component in association with changes in scale) are discussed also.

symmetry operations

As indicated earlier there are four distinct symmetry operations of relevance to two-dimensional designs and their classification: rotation, reflection, translation and glide reflection. Each is described in this section.

Rotation allows repetition of a figure or a component of a design at regular intervals round an imaginary point (known as a centre of rotation). Any figure exhibiting regular circle-wise repetition is deemed to have rotational symmetry. Rotational symmetry is classified in terms of divisions of 360 degrees. So if a figure exhibits a component part which repeats at 60-degree intervals around a centre of rotation, then that motif is deemed to have six-fold rotational symmetry. The order of rotational symmetry of a figure is the number of times that a fundamental component part repeats itself in one revolution (of 360 degrees) around the centre of rotation. So a figure exhibiting six-fold rotational symmetry has six equal parts distributed at equal distances around the centre.

Reflection allows a figure to undergo repetition across an imaginary line, known as a reflection axis, producing a mirror image, which may be denoted in illustrative form by the letter m (from the English *mirror*). So in this case a figure will consist of two equal parts (or fundamental units) after reflection in a reflection axis. Reflection across a single reflection axis is characteristic of so-called bilateral symmetry, where two-dimensional figures or other objects are considered to be divided into two equal mirror-image parts, a characteristic prevalent among humans and the majority of animals (Shubnikov and Koptsik 1974: 11). Reflection can be a characteristic of motifs (considered in 'motifs, figures or repeating units'), frieze patterns (considered in 'friezes, strips or borders') and all-over patterns (considered in 'lattices and all-over patterns'). It is worth noting that where a component (i.e. the fundamental unit) of a motif or pattern has a directional characteristic (i.e. with clockwise or anti-clockwise orientation, such as an *S* shape or a *Z* shape, respectively), this orientation will be reversed on reflection, as is the case when viewing such a reflection in a conventional mirror (so reflections reverse directions).

Translation is the simplest symmetry operation (shown previously in Figure 5.1). It allows a motif (i.e. the repeating unit within a regularly repeating design) to undergo repetition vertically, horizontally or diagonally at regular intervals while retaining the same orientation (Coxeter 1961: 34). Abas and Salman, in their treatise dealing with symmetry in Islamic tilings, defined a translation as 'a movement which causes every point in an object to shift by the same amount in the same direction' (1995: 56). Translation in one consistent direction results in a frieze pattern and, if carried out in two independent directions across the plane, yields an all-over pattern.

The fourth symmetry operation, glide reflection (shown previously in Figure 5.1), allows a motif to be repeated in one action through a combination of translation and reflection, in association with a glide-reflection axis. A frequently cited example in explanatory texts is the impression created by a person's or an animal's footprints, each impression at a regular distance from the previous impression and each a reflected image of the previous.

motifs, figures or repeating units

From the viewpoint of symmetry, it is common to classify motifs (or figures) employing a series of symbols. Using the letter n to represent any whole number, a motif may be considered to have n-fold rotational symmetry about a fixed point, when it can be seen that a component of the motif is repeated by successive rotations of 360/n degrees about a fixed point. After n successive rotations (each of 360/n degrees), the component is returned to its original position within the motif. Such motifs may be classified using the notation cn; in this case, the letter 'c' denotes circle-wise and 'n' any whole number associated with the number of stages of rotation within 360 degrees. It should be mentioned also that some motifs are asymmetrical: these have no symmetrical properties, and

their constituent elements can coincide only with themselves after a full rotation of 360 degrees. Thus It can be seen that

n = 1 for rotations of 360 degrees (i.e. asymmetrical motifs);

n = 2 for rotations of 180 degrees (i.e. two-fold rotation);

n = 3 for rotations of 120 degrees (i.e. three-fold rotation);

n = 4 for rotations of 90 degrees (i.e. four-fold rotation);

n = 5 for rotations of 72 degrees (i.e. five-fold rotation);

n = 6 for rotations of 60 degrees (i.e. six-fold rotation).

Each of these relationships is shown schematically in Figure 5.2. The smallest element of a motif which can be rotated by an angle of 360/n degrees is known as the fundamental unit of that motif. The number of times that the fundamental unit comes into coincidence with itself (or overlaps exactly with itself) during one full rotation corresponds to the order of rotation (i.e. two-fold, three-fold etc.). Motifs may exhibit higher orders of rotation than those indicated above (e.g. c7, c8, c9, c10 etc.), depending on the number of times that the fundamental unit of the motif is repeated (or coincides with itself) around a centre of rotation in a 360-degree turn.

Class c1 motifs are asymmetrical, do not have equal parts and so must rotate by 360 degrees for constituent parts to coincide (Figure 5.3).

Class c2 motifs possess only two-fold rotational symmetry and consist of two fundamental units. Through rotation of 180 degrees, each fundamental unit will come into coincidence with its neighbour. With a further 180-degree rotation,

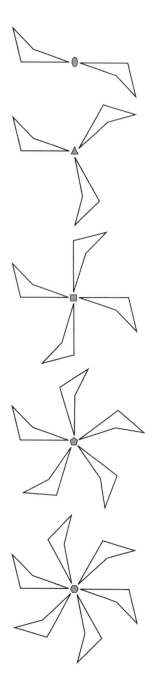

Figure 5.3 A class c1 motif (JSS, from Hann, 1991).

Figure 5.2 Schematic illustrations of class cn motifs (JSS).

each unit will be placed in exactly its original position. As a result, each point on each fundamental unit will have its equivalent point in its neighbouring

unit. Examples are shown in Figure 5.4a–b. In each case the fundamental unit is half the area of the motif.

Class c3 motifs exhibit three-fold rotational symmetry, with rotations of 120 degrees, 240 degrees and 360 degrees bringing the fundamental unit of the motif into coincidence at each turn. Examples are shown in Figure 5.5a–b.

Figure 5.6a–b shows examples of motifs from class c4, each with a minimum angle of rotation of 90 degrees. These motifs are therefore characterized by the presence of rotations through 90 degrees, 180 degrees, 270 degrees and 360 degrees.

Class c5 motifs (Figure 5.7a–b) have a minimum rotational angle of 72 degrees and, when rotated progressively by this amount, the fundamental unit comes into coincidence with itself five times.

Six-fold rotation characterizes c6 motifs, examples of which are shown in Figure 5.8a–b. During a complete rotation of 360 degrees the

a

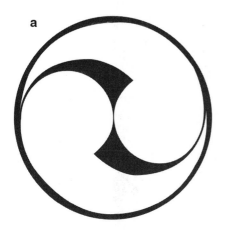

b

Figure 5.4a–b Class c2 motifs (MV, developed from Hann, 1991).

a

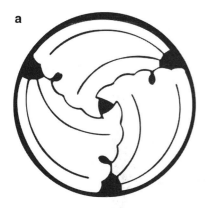

b

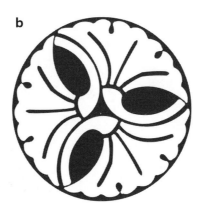

Figure 5.5a–b Class c3 motifs (MV, developed from Hann, 1991).

a

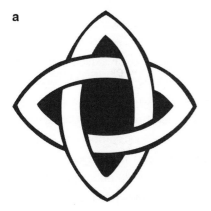

b

Figure 5.6a–b Class c4 motifs (MV, developed from Hann, 1991).

a

b

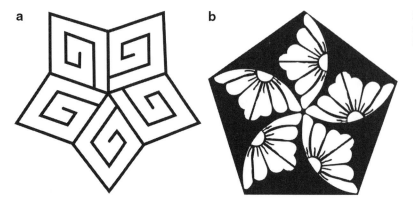

Figure 5.7a–b Class c5
motifs (MV, developed from
Hann, 1991).

a

b

Figure 5.8a–b Class c6
motifs (MV, developed from
Hann, 1991).

fundamental unit of the motif will coincide with it-self six times (since the minimum angle of rotation is 60 degrees).

Rotational symmetry on its own characterizes one class of symmetrical motifs, and a combination of reflectional and rotational symmetry characterizes the other class. The order of reflectional symmetry of a motif is equal to the number of reflection axes passing through and intersecting (if more than one) through the motif's centre. Motifs with reflectional symmetry are classed as dn motifs ('d' for dihedral, indicating two component parts, and 'n' equal to any whole number). Schematic representations of various dn motifs

are given in Figure 5.9. A d1 motif has one reflection axis passing through its centre and thus has two component equal parts, one a reflection of the other as if in a mirror (Figure 5.10a–b). This is characteristic of so-called bilateral symmetry (mentioned earlier); these motifs are the only motifs within this class which exhibit reflectional symmetry and no rotational symmetry. Class d2 motifs (Figure 5.11a–b) have bilateral symmetry around both their horizontal and their vertical axes, so there are four component parts (or fundamental units). Each motif has two reflection axes, intersecting at 90 degrees. The motif may be produced also by rotating a bilaterally

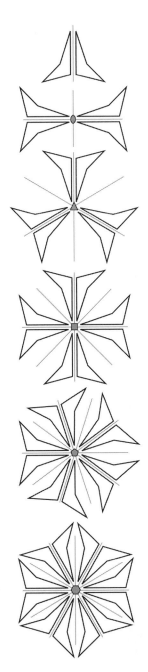

Figure 5.9 Schematic illustration of dn motifs (JSS).

but also a secondary symmetry characteristic of rotation. Conceptually, it is most convenient to consider the rotational properties of motifs of orders d2 and above as byproducts of successive reflection operations rather than as the primary symmetry constituents.

Class d3 motifs (Figure 5.12a–b) are characterized by the presence of three intersecting reflection axes which produce bilaterally symmetrical units spaced at 120 degrees. The fundamental unit is contained in one-sixth of a circle. This class of motifs may also be produced by rotations of a bilaterally symmetrical unit through 120 degrees, 240 degrees and 360 degrees.

Class d4 motifs (Figure 5.13a–b), are characterized by the presence of four-fold rotational symmetry together with four intersecting reflection axes (which pass through the centre of four-fold rotation and are intersected at an angle of 45 degrees). The fundamental unit takes up one-eighth of the total motif area and, when reflected, will produce a bilaterally symmetrical unit which may be rotated in 90-degree stages to produce the whole motif.

Class d5 motifs (Figure 5.14a–b) are characterized by the presence of five reflection axes and five-fold rotational symmetry. The fundamental unit, which takes up one-tenth of the total motif area, may be reflected to produce a bilaterally symmetrical unit which when rotated five times produces the full motif.

Six intersecting reflection axes characterize d6 motifs. Typical examples are snowflakes (Figure 5.15a–b). The fundamental unit takes up one-twelfth of the motif's area. Motifs from this class may be produced also by rotations of a bilaterally symmetrical unit through 60 degrees, 120 degrees, 180 degrees, 240 degrees, 300 degrees and 360 degrees.

symmetrical unit (i.e. one half of the motif) through 180 degrees. So where two or more reflection axes are present, motifs exhibit not only reflection

a
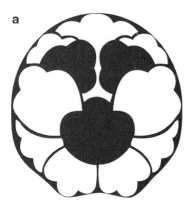

b
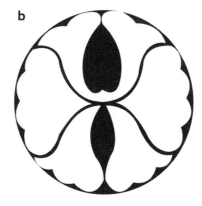

Figure 5.10a–b Class d1 motifs (MV, developed from Hann, 1991).

a
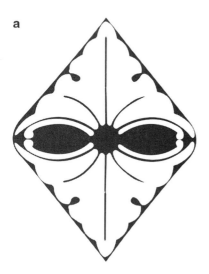

b

Figure 5.11a–b Class d2 motifs (MV, developed from Hann, 1991).

a
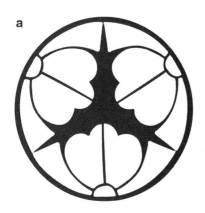

b
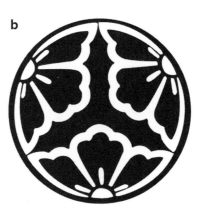

Figure 5.12a–b Class d3 motifs (MV, developed from Hann, 1991).

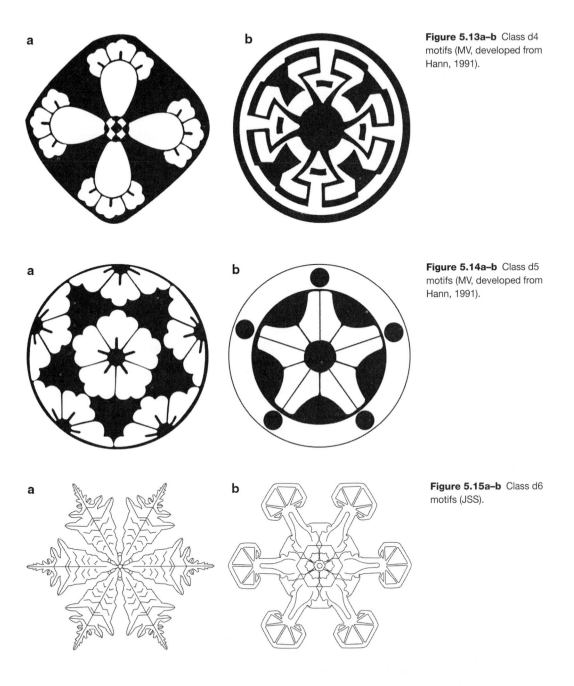

a   b

Figure 5.13a–b Class d4 motifs (MV, developed from Hann, 1991).

a   b

Figure 5.14a–b Class d5 motifs (MV, developed from Hann, 1991).

a   b

Figure 5.15a–b Class d6 motifs (JSS).

friezes, strips or borders

As noted in the 'symmetry operations' section, the four symmetry operations or actions are as follows: rotation, by which a motif or a component of a motif is rotated about a fixed point so that it undergoes repetition at regular angular intervals (imagined within a circle); reflection, by which a motif or a component of a motif is reflected across a straight line (or reflection axis) producing a mirror image characteristic of so-called bilateral symmetry; translation, or repetition at regular intervals of a motif in a straight line without change of orientation; glide reflection, by which a motif is repeated using a combination of translation and reflection.

Various scholars have investigated, explained and discussed the different means of combining these four symmetry operations to create designs which translate (repeat) in one direction (i.e. between two parallel lines) to produce what are known as frieze (strip or border) patterns, or in two directions (i.e. across the plane) to produce all-over patterns. With both pattern types, a constituent part of the design repeats regularly without change in its shape, size, orientation or content. When symmetry operations are combined in the form of frieze patterns, only seven distinct

combinations are possible. As indicated by the title of this section, the objective is to examine further the symmetry combinations possible with frieze patterns. Certain restrictions apply. A pattern's rotational characteristics must be applicable across or along the whole pattern. It is worth noting at this stage that only rotations of the order of two (180 degrees) are possible in frieze patterns. This is because all elements need to be superimposed along the length of the pattern at each stage of rotation, and only two-fold rotation would allow this re-positioning to occur between the two (imaginary) parallel lines which constitute a frieze pattern's spatial boundaries. Schematic illustrations are provided in Figure 5.16.

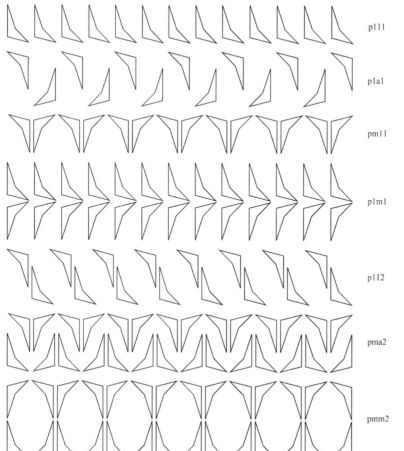

p111

p1a1

pm11

p1m1

p112

pma2

pmm2

Figure 5.16 Schematic illustrations of the seven frieze patterns (JSS).

Unfortunately, a range of different notations has been used by various authors, and the most commonly accepted notation is not straightforward. It comes in the form of four symbols (generally denoted by pxyz), where each of the last three symbols denotes the presence of a particular symmetry operation within the border pattern. The letter p denotes that the design lies on the plane. The seven frieze pattern types are as follows: p111, p1a1, pm11, p1m1, p112, pma2, pmm2.

Washburn and Crowe (1988) gave a relatively straightforward explanation of the notation. The first symbol ('p') prefaces all seven classes of border patterns. Symbols placed in the second, third and fourth positions indicate the presence of vertical reflection, horizontal reflection or glide reflection, and two-fold rotation respectively. A fuller explanation of the notation is given in Appendix 2.

Class p111 frieze patterns

From the viewpoint of symmetry, this is the most elementary of the frieze patterns since the only operation that the pattern possesses is translation (Figure 5.17a–b).

Class p1a1 frieze patterns

Glide reflection, as explained previously, is a combination of translation followed by a reflection in a line parallel to the translation axis of the frieze pattern (Figure 5.18a–b).

Class pm11 frieze patterns

Class pm11 frieze patterns are characterized by the presence of translation combined with reflections across two alternating vertical reflection axes, each perpendicular to the central axis of the frieze pattern (Figure 5.19a–b).

Class p1m1 frieze patterns

Class p1m1 frieze patterns have reflection in a longitudinal axis through the centre line of the frieze, combined with translation along the axis of the pattern (Figure 5.20a–b).

Class p112 frieze patterns

Class p112 frieze patterns exhibit two-fold rotational symmetry, combined with translation along the axis of the pattern (Figure 5.21a–b).

Figure 5.17a–b Class p111 frieze patterns (MV, developed from Hann, 1991).

a

b

Figure 5.18a–b Class p1a1 frieze patterns (MV, developed from Hann, 1991).

a

b

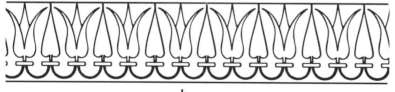

Figure 5.19a–b Class pm11 frieze patterns (MV, developed from Hann, 1991).

a

b

Figure 5.20a–b Class p1m1 frieze patterns (MV, developed from Hann, 1991).

a

b

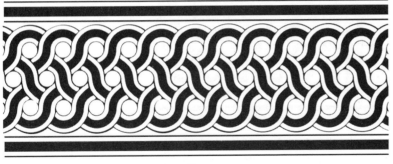

Figure 5.21a–b Class p112 frieze patterns (MV, developed from Hann, 1991).

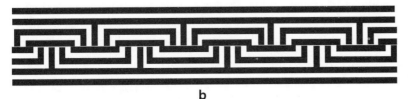

Figure 5.22a–b Class pma2 frieze patterns (MV, developed from Hann, 1991).

Class pma2 frieze patterns

Class pma2 frieze patterns (Figure 5.22a–b) combine all four symmetry operations.

Class pmm2 frieze patterns

Class pmm2 frieze patterns (Figure 5.23a–b) have a horizontal reflection axis which is intersected at regular points by perpendicular reflection axes.

Two-fold rotation centres are thus established in the intersection points of the axes.

lattices and all-over patterns

Before explaining the symmetry characteristics of all-over patterns, it is worth noting that all such

a

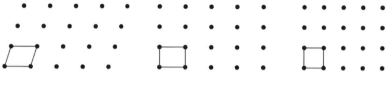

b

Figure 5.23a–b Class pmm2 frieze patterns (MV, developed from Hann, 1991).

patterns are built on frameworks known as lattices. All patterns have sets of corresponding points (equal points on identical positions on motifs oriented in the same direction) which may be connected to form regular frameworks or lattice structures (of which there are five varieties). These lattice structures consist of unit cells of identical size, shape and content which, when translated across the lattice structure, in two independent, non-parallel directions, produce the full repeating pattern. Combinations of the four symmetry operations yield a total of seventeen classes of all-over

patterns. Each of the seventeen is associated with one of five lattice structures, and each lattice structure has associated with it a particular shape of parallelogram (or unit cell) which contains the repeating unit of the design. The five lattices are known as Bravais lattices, and the unit cells associated with each are shown in Figure 5.24 (clockwise from top left): parallelogram, rectangle (adjacent sides not equal and all internal angles of 90 degrees), square (all sides equal and each internal angle 90 degrees), hexagon (by connecting six points around one) and rhombic diamond

Figure 5.24 The five Bravais lattices (MV, developed from Hann, 1991).

(sides of equal length). Schematic illustrations of unit cells for each of the seventeen all-over pattern classes are given in Figure 5.25. It can be seen that the rhombic lattice unit, unlike the other lattice units, is centred and has a diamond-shaped cell held within a rectangle (denoted by dashed lines) so that reflection axes can be positioned at right angles to the sides of the enlarged cell, which holds one full repeating unit (within the diamond-shaped cell) and a quarter of a repeating unit at each of the enlarged cell corners.

As with frieze patterns, the seventeen all-over pattern classes have various notations associated with them. The situation is more complex than with border patterns as the relevant symmetry operations operate fully across the plane and not just between two imaginary parallel lines. As indicated earlier, there are also seventeen alternative symmetry combinations (or classes) rather than seven. The least complicated notation has been selected for use here. In summary, it identifies the highest order of rotation within the pattern together with the presence (or absence) of glide reflection and/or reflection. Where rotational symmetry is present, only two-, three-, four- and six-fold orders of rotation (and combinations of these)

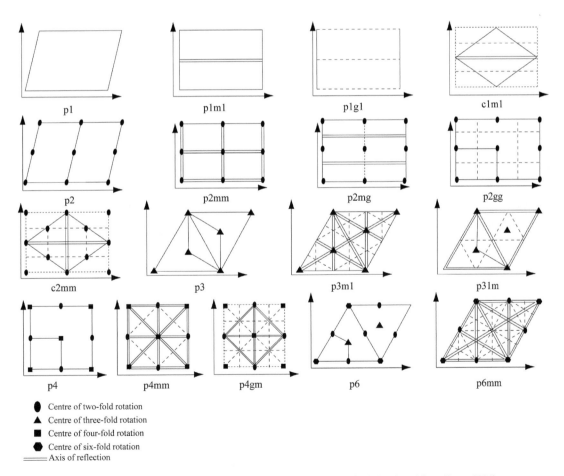

Figure 5.25 Unit cells for each of the seventeen classes of all-over patterns (MV, developed from Hann, 1991).

are possible in the production of regularly repeating all-over patterns, as figures with five-fold rotational symmetry cannot repeat themselves around a 360 degree axis. This is known as the crystallographic restriction and was discussed by Stevens (1984: 376–90). Reflection, where it is present, may be in one or more directions, and may combine with other symmetry operations; the same is true of glide reflection. Schematic illustrations of the seventeen all-over pattern classes are provided in Figure 5.26, and a fuller explanation of the notation is given in Appendix 2. A good introduction to the subject is that by Padwick and Walker (1977).

In describing the seventeen classes of all-over patterns, it is best to group them by making reference to the absence or presence of rotational symmetry (and, if present, the highest order of rotational symmetry): pattern classes p111, p1g1, p1m1 and c1m1 do not include rotation of any kind among their constituent symmetries; pattern classes p211, p2gg, p2mg, p2mm and c2mm exhibit two-fold rotation; classes p311, p3m1 and p31m exhibit three-fold rotational symmetry; classes p411, p4gm and p4mm exhibit four-fold rotation; classes p611 and p6mm have a highest order of rotation of six. It should be noted that different orders of rotation combine in certain all-over pattern classes, and the highest constituent order of rotation determines the grouping here.

Patterns without rotational symmetry

From the viewpoint of pattern symmetry, class p111 (shortened to class p1) all-over patterns are the most straightforward in terms of analysis, classification and construction. The parallelogram unit cell is the one chosen conventionally; however, since the pattern has no constituent symmetries other than translation, any of the five lattice types and their associated cells would be suited. So this pattern type does not exhibit rotation, reflection or glide reflection (Figure 5.27a–b).

Glide reflections and translations are the constituent symmetry characteristics of class p1g1 (shortened to class pg) all-over patterns (Figure 5.28a–b).

Class p1m1(shortened to class pm) all-over patterns are characterized by reflections and translations (Figure 5.29a–b).

Class c1m1 (shortened to class cm) all-over patterns are characterized by a unit cell of the rhombic lattice type which contains a diamond-shaped cell held (or centred) within a larger rectangle. Since the repeating cell is centred, the letter c prefaces the notation. Constituent symmetry characteristics include translations combined with reflections and glide reflections (Figure 5.30a–b).

Two-fold rotational symmetry

Class p211 (shortened to class p2) all-over patterns contain translations combined with two-fold rotations (Figure 5.31a–b).

Class p2gg (shortened to class pgg) all-over patterns contain translations combined with glide reflections and two-fold rotations (Figure 5.32a–b).

Class p2mg (shortened to class pmg) all-over patterns combine translations with glide reflections, reflections and two-fold rotations (Figure 5.33a–b).

Class p2mm (shortened to class pmm) all-over patterns combine translations with reflections in two directions and two-fold rotations (Figure 5.34a–b).

Class c2mm (shortened to class cmm) all-over patterns are built on a centred cell (thus the preface c) and include parallel-reflection and glide-reflection axes, in both vertical and horizontal

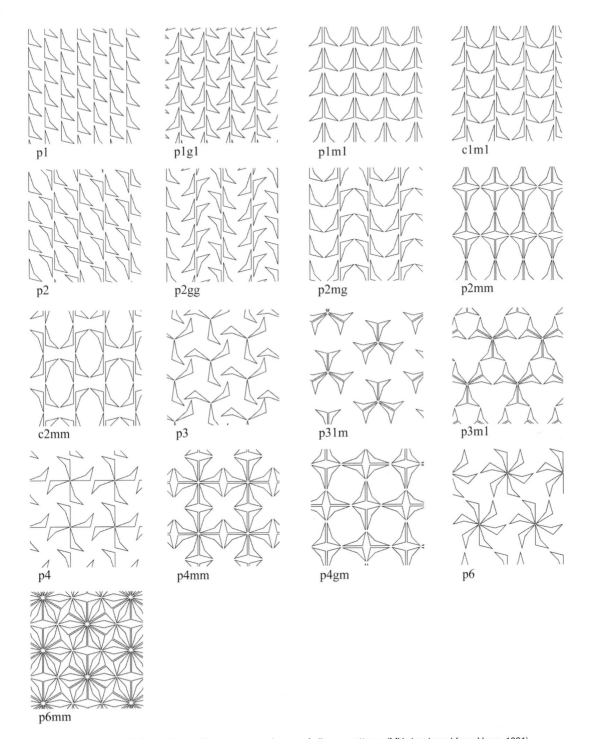

Figure 5.26 Schematic illustrations of the seventeen classes of all-over patterns (MV, developed from Hann, 1991).

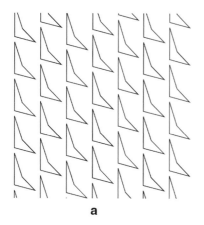

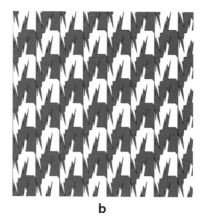

a   b

Figure 5.27a–b Class p111 all-over patterns (JSS and MV, developed from Hann, 1991).

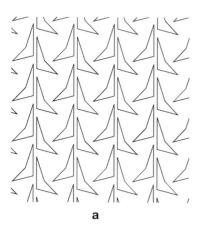

a   b

Figure 5.28a–b Class p1g1 all-over patterns (JSS and MV, developed from Hann, 1991).

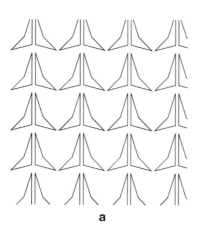

a   b

Figure 5.29a–b Class p1m1 all-over patterns (JSS and MV, developed from Hann, 1991).

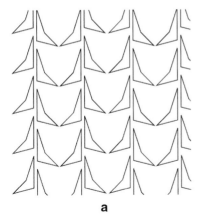

a

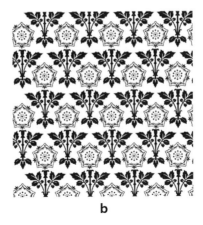

b

Figure 5.30a–b Class c1m1 all-over patterns (JSS and MV, developed from Hann, 1991).

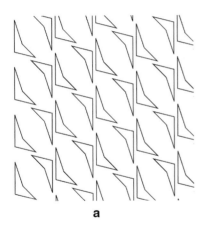

a

b

Figure 5.31a–b Class p211 all-over patterns (JSS and MV, developed from Hann, 1991).

a

b

Figure 5.32a–b Class p2gg all-over patterns (JSS and MV, developed from Hann, 1991).

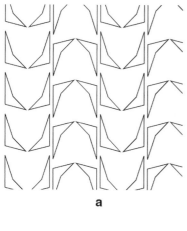

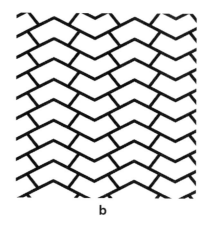

a   b

Figure 5.33a–b Class p2mg all-over patterns (JSS and MV, developed from Hann, 1991).

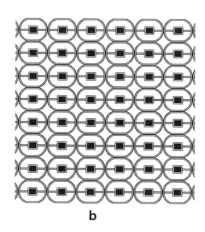

a   b

Figure 5.34a–b Class p2mm all-over patterns (JSS and MV, developed from Hann, 1991).

directions, combined with translations and two-fold rotations (Figure 5.35a–b).

Three-fold rotational symmetry

Class p311(shortened to class p3) all-over patterns combine translations with three-fold rotations (Figure 5.36a–b).

Class p3m1 all-over patterns combine translations with three-fold rotations and reflections (Figure 5.37a–b). Each three-fold rotational centre is positioned at the intersection of reflection axes (which differentiates this class from class p31m).

Class p31m all-over patterns combine translations with three-fold rotations and reflections (as with the previous class). With class p31m, not all three-fold rotational centres are on reflection axes (Figure 5.38a–b).

Four-fold rotational symmetry

Class p411 (shortened to class p4) all-over patterns combine translations with a highest order of rotation of four, and points also of two-fold rotation (Figure 5.39a–b).

Class p4gm (shortened to class p4g) all-over patterns combine translations with a highest order of rotation of four, together with two-fold rotations, reflections in two directions and glide reflections (Figure 5.40a–b).

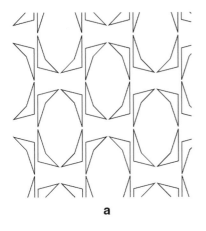

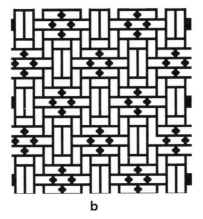

Figure 5.35a–b Class c2mm all-over patterns (JSS and MV, developed from Hann, 1991).

a   b

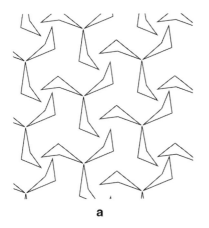

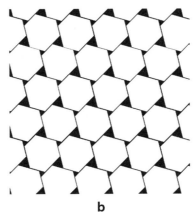

Figure 5.36a–b Class p311 all-over patterns (JSS and MV, developed from Hann, 1991).

a   b

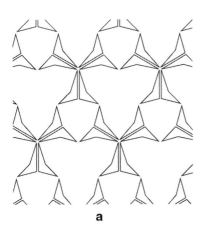

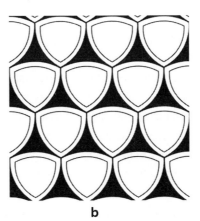

Figure 5.37a–b Class p3m1 all-over patterns (JSS and MV, developed from Hann, 1991).

a   b

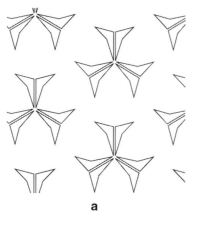

a   b

Figure 5.38a–b Class p31m all-over patterns (JSS and MV, developed from Hann, 1991).

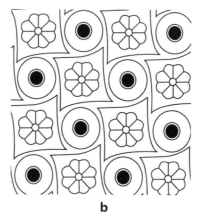

a   b

Figure 5.39a–b Class p411 all-over patterns (JSS and MV, developed from Hann, 1991).

Class p4mm (shortened to class p4m) all-over patterns combine translations with a highest order of rotation of four, as well as two-fold rotation and reflection axes in horizontal, vertical and two diagonal directions (Figure 5.41a–b).

Six-fold rotational symmetry

Class p611 (shortened to class p6) all-over patterns combine translations with a highest order of rotation of six and include also points of three-fold and two-fold rotation (Figure 5.42a–b).

Class p6mm(shortened to class p6m) all-over patterns combine translations with a highest order of rotation of six points of both three-fold

and two-fold rotation and reflections in several directions (Figure 5.43a–b).

counter-change designs

Systematic colour change in motifs and in regular patterns can be explained also in terms of symmetry concepts. In many repeating designs, colour may be reproduced simply with no distributional change from one repeating unit to the next. Colour is thus preserved and the same component of each repeating unit is coloured identically across the design. This has been the case with all patterns illustrated so far in this chapter. Colour may be changed

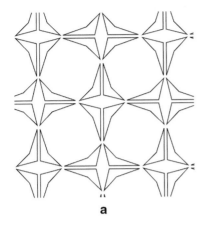

a

b

Figure 5.40a–b Class p4gm all-over patterns (JSS and MV, developed from Hann, 1991).

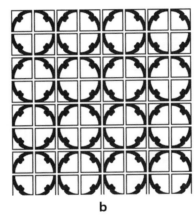

a

b

Figure 5.41a–b Class p4mm all-over patterns (JSS and MV, developed from Hann, 1991).

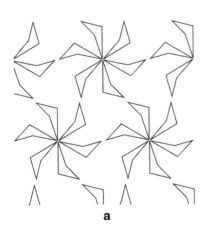

a

b

Figure 5.42a–b Class p611 all-over patterns (JSS and MV, developed from Hann, 1991).

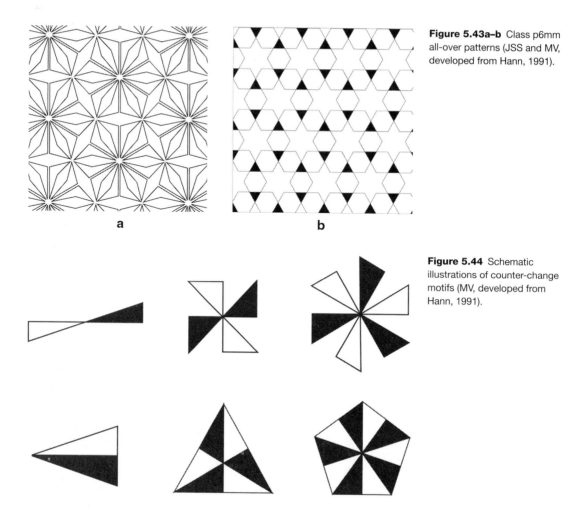

Figure 5.43a–b Class p6mm all-over patterns (JSS and MV, developed from Hann, 1991).

Figure 5.44 Schematic illustrations of counter-change motifs (MV, developed from Hann, 1991).

a   b

systematically, however, to produce what are known as counter-change designs (Figure 5.44), a term used by Christie (1910, and Ch.10 of 1969 edition), Gombrich (1979: 89) and Woods (1936). Examples created by the artist Ihab Hanafy, when asked to respond to the theme 'counter-change in art and design', are presented in Figures 5.45–5.50.

The relevant principles governing colour counter-change in regularly repeating patterns, involving two-, three- and higher-coloured symmetries, were explained and illustrated by Hann (2003a, b, c) and Hann and Thomas (2007). Although the concern of this section has been with

systematically changing colours in a regularly repeating pattern, there is no reason why other characteristics such as textures, raw materials used and physical properties (of whatever kind) cannot be counter-changed in the same way.

fractals and self-similarity— another type of symmetry

A fractal is a geometrical shape which exhibits a particular type of repetition and is made up of identical parts, each of which is (at least

Figure 5.45 'Counter-change 1' (Ihab Hanafy).

Figure 5.47 'Counter-change 3' (Ihab Hanafy).

Figure 5.46 'Counter-change 2' (Ihab Hanafy).

Figure 5.48 'Counter-change 4' (Ihab Hanafy).

approximately) a reduced-size copy of the whole. The repetitive action is known as iteration, and the type of symmetry is known as scale symmetry (or self-similarity). The term *fractal* was coined by Benoît Mandelbrot to describe such an object or figure. These figures exhibit high degrees of self-similarity. A well-known example of a fractal image is the Cantor Bar Set (named after the nineteenth-century German mathematician Georg Cantor). This may be constructed by dividing a line into

Figure 5.49 'Counter-change 5' (Ihab Hanafy).

Figure 5.50 'Counter-change 6' (Ihab Hanafy).

three equal parts and removing the middle part (Figure 5.51). The procedure is repeated indefinitely, first on the two remaining parts, then on the four parts produced by that operation, and so on, until the resultant object has an infinitely large number of parts each of which is infinitely small.

Fractals can be based on mathematical models but are also common in real life. Examples of nature's fractals are various vegetables (e.g. cauliflower and broccoli), clouds, lightening, trees, coastlines and mountains. In the built environment, fractals can be detected in the exterior and interior of Gothic cathedrals, where typically arch-shaped features are presented at a range of scales. *Basilica y Templo Expiatorio de la Sagrada Familia*, commonly known as Sagrada Familia, designed by Antoni Gaudi (1852–1926), also expresses self-similarity (Figure 5.52a–b). Bovill (1996) offers a good source from which to develop your knowledge of the subject. A series of images produced by students as a response to the theme 'fractals' is presented in Figures 5.53–5.66. Images produced by Robert Fathauer, a renowned expert in the field of fractal tessellations, are presented in Figures 5.67–5.69. Figure 5.70 shows an image produced by Craig S. Kaplan, an accomplished researcher, scholar and practitioner.

Figure 5.51 Cantor Bar Set (JSS).

a

b

Figure 5.52a–b Sagrada Familia, designed by Antoni Gaudi, Barcelona, 2007.

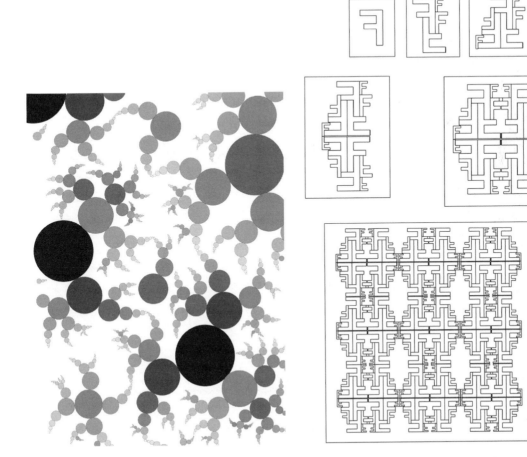

Figure 5.53 Fractal image (Olivia Judge).

Figure 5.54 Fractal images (Nathalie Ward).

Figure 5.55 Fractal images (Reerang Song).

Figure 5.57 Fractal images (Natasha Purnell).

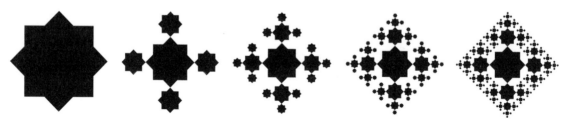

Figure 5.56 Fractal images (Alice Simpson).

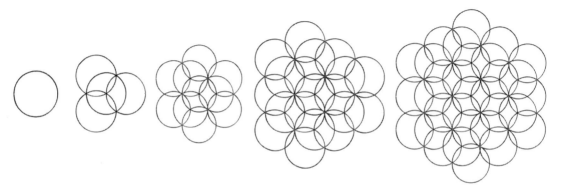

Figure 5.58 Fractal images (Esther Oakley).

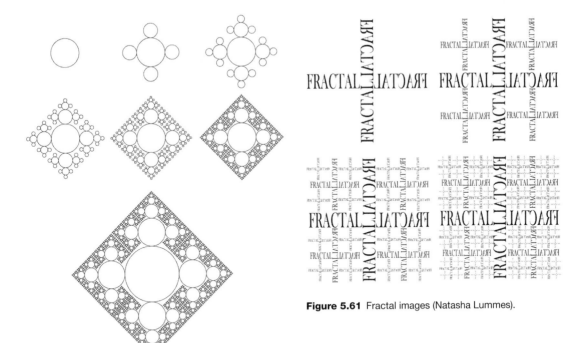

Figure 5.61 Fractal images (Natasha Lummes).

Figure 5.59 Fractal images (Josef Murgatroyd).

Figure 5.60 Fractal images (Robbie Macdonald).

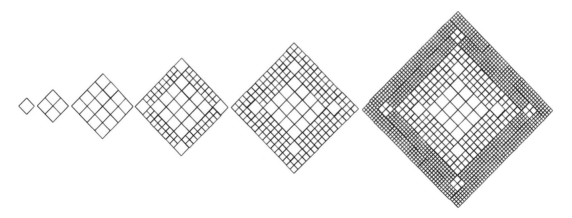

Figure 5.62 Fractal images (Rachel Lee).

Figure 5.63 Fractal images (Edward Jackson).

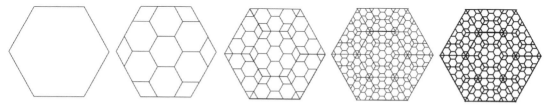

Figure 5.64 Fractal images (Zanhib Hussain).

Figure 5.65 Fractal images (Jessica Dale).

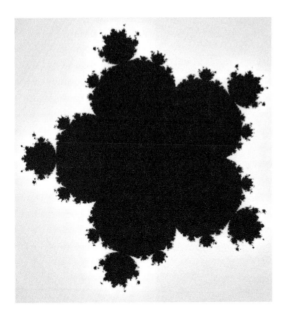

Figure 5.66 Fractal image (Daniel Fischer).

Figure 5.68 'Fractal Tree No. 5', by Robert Fathauer, is a digital artwork constructed in 2007 by graphically iterating a photographic building block created from photographs of the base of a rose bush. The rugged appearance of the thorny stalks is juxtaposed with the light, feathery appearance of the overall fractal.

Figure 5.67 'Fractal Tessellation of Spirals'. Robert Fathauer is a graduate in physics and mathematics and was awarded a PhD in electrical engineering. He began designing his own tessellations in the late 1980s. A decade later his interests branched into the area of fractals as well as fractal knots and fractal trees. He has published and exhibited widely. 'Fractal Tessellation of Spirals' is a digital artwork completed in February 2011. It is based on a fractal tessellation of kite-shaped tiles discovered by Dr. Fathauer a decade previously. All the spirals in the print have the same shape (more precisely, they are all similar in the Euclidean plane).

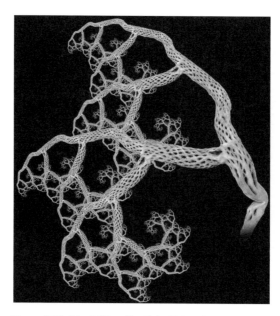

Figure 5.69 'Fractal Tree No. 7', by Robert Fathauer, is a digital artwork constructed in 2009 by graphically iterating a photographic building block created from photographs of the skeleton of a cholla cactus.

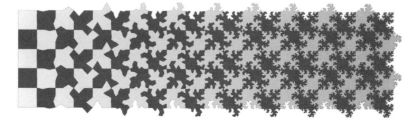

Figure 5.70 'Parquet deformation', illustration by Craig S. Kaplan. The design is a 'parquet deformation', inspired by the design studio of William Huff, in which a tessellation of the plane is constructed from shapes that evolve slowly in space. Here, squares evolve into shapes with fractal boundaries.

CHAPTER SUMMARY

Concepts associated with symmetry have been used to describe and classify designs, particularly regularly repeating designs, typically characterized by the systematic (or regular) repetition of a tile, motif or figure along a band (known as a frieze pattern) or across the plane (known as an all-over pattern). The use of the term *pattern* denotes the perceived presence of order. This chapter recognizes the all-encompassing nature of symmetry and, in particular, how it can be used to characterize two-dimensional repeating designs (or patterns) and their component parts.

It has been shown that motifs may be classified into one of two general classes dependent upon the symmetry operations used in their construction. Class cn motifs, of an order higher than c1, exhibit rotational characteristics only, and class dn motifs, of an order higher than d1, exhibit combinations of rotation and reflection. The symmetry operations of translation, rotation, reflection and glide reflection may be combined to produce a total of seven (and only seven) classes of regularly repeating frieze patterns; when used in combination with five lattice varieties (known as Bravais lattices), seventeen classes of all-over patterns are possible. Patterns may be classified by reference to their symmetry characteristics: if two patterns have the same symmetry characteristics they are thus of the same symmetry class.

Systematic colour change in motifs and in regular patterns and tilings can be explained also in terms of symmetry concepts. In many repeating designs, repeating units and their parts remain the same colour or colours throughout. Colour may be changed systematically and in a continuous way to produce what are known as counter-change designs.

A fractal is a geometric shape made up of identical parts, each of which is (at least approximately) a reduced-size copy of the whole. Although fractals were known since the late-nineteenth century (when they were regarded as mathematical curiosities), their true nature was revealed in the 1960s and 1970s through the studies of Benoît Mandelbrot and other scientists. The term *fractal* itself was coined by Mandelbrot to describe a complex geometrical object that has a high degree of self-similarity.

6

the stepping stones of Fibonacci and the harmony of a line divided

introduction

In the treatise *Mysterium Cosmographicum* (1596), Kepler (1571–1630 CE), the great astronomer, proclaimed that geometry had two great treasures: one was the Pythagorean theorem, and the other was a line divided into two unequal lengths such that the length of the original line is to the longer of the two cut lengths as the longer cut length is to the shorter cut length (known to geometers as a line divided into extreme and mean ratio). The first treasure he likened to a measure of gold and the second to a 'precious jewel' (Huntley 1970: 23, and Fletcher 2006). It is this second measure and related geometrical phenomena often deemed to be of value to artists and designers that is the concern of this chapter.

Early thinking relating to many of the principles, figures and objects of geometry dealt with in this book can be associated invariably with early Greek mathematicians such as Euclid (325–265 BCE). The so-called golden section is one such construction. A definition was provided by Euclid (325–265 BCE) in his *Elements:* 'A straight line is said to have been cut in extreme and mean ratio when, as the whole line is to the greater segment, so is the greater to the less' (Heath 1956: 188). All art-and-design practitioners and teachers have probably heard of the golden section, but many

have, in all probability, no real understanding of what it actually is or how it could be applied in the creative process. So this chapter is focused on remedying this state of affairs.

It is worth remarking at this stage that the term *golden section* may be rather confusing, because the word 'section' in everyday English means 'part'. However, in the expression 'golden section', the word 'section' means 'cutting', 'dividing' or, perhaps better, 'point of division'. The crucial aspect is that the golden section is a ratio of one length to another (expressed as approximating to 1.6180:1). It seems that the earliest treatise dealing with the golden section itself is *De Divina Proportione* (1509) by Luca Pacioli (1445–1517 CE) (Huntley 1970: 25, and Olsen 2006: 2). This work was illustrated by Leonardo da Vinci, who, tradition claims, used the term *sectio aurea,* or golden section, though it seems that the first published use of the term is in Martin Ohm's 1835 treatise, *Pure Elementary Mathematics* (Olsen 2006: 2).

Through the years, numerous other terms have been coined, including golden or divine ratio, mean, number, cut or proportion. In the twentieth century, Mark Barr, an American mathematician, proposed the use of the Greek letter Φ (phi) to denote a numerical value which approximates to 1.6180, a value which, as indicated earlier, is closely associated with the golden section.

Phi can be derived in many ways and is reputed to show up in relationships throughout the natural, constructed and manufactured worlds; it has been associated with the dimensions of various famous buildings, the proportions of the human body, other animals, plants, DNA, the solar system, music, dance, painting, sculpture and other art forms. However it should be stressed that there are contrary views on the past use of the golden section by artists, architects, builders and designers, and many eminent scholars have presented convincing reviews identifying some of the misconceptions which have built up in the literature over the years.

This chapter introduces the Fibonacci series, explains and illustrates the various golden-section-related constructions and presents a summary of the debate relating to the detection of the golden section in art and architecture.

the Fibonacci series

A particular numerical series (ultimately named the Fibonacci series) was introduced into Europe by Leonardo of Pisa (ca.1170–1250 CE). Fibonacci, as he was known, had spent much of his early life in North Africa where he learned the Hindu-Arabic numerals and the associated decimal system. Subsequently to returning to Italy in his early thirties, he published a book in 1202 under the title *Liber Abaci* (the *Book of the Abacus* or *Book of Calculation*) and thus helped introduce Hindu-Arabic numerals to Europe. Presented in this book was a problem relating to the progeny of a single idealized (biologically unrealistic) pair of rabbits, one male and one female. The problem supposed that a pair of newly born rabbits was placed in an enclosed field in January of a given

year. In February they mated and produced a second pair in the month of March; thereafter, each pair of rabbits (including the first pair) produced another pair, in the second month after birth and, subsequently, one pair each month (Huntley1970: 158). The problem was to establish the total number of pairs by the end of the year. Fibonacci showed that the successive monthly numbers of pairs of rabbits conformed to the numerical series 1, 1, 2, 3, 5, 8, 13, 21, 34, 55... (Figure 6.1). So, at the end of December in the given year, the total population was 144 pairs of rabbits. Although he is credited with inventing this numerical sequence, Fibonacci probably acquired relevant knowledge from Islamic scholars in North Africa; these scholars in turn probably acquired knowledge of the numerical sequence from Indian mathematicians. Apparently, the Fibonacci series was known to Indian mathematicians as early as the sixth century (Goonatilake 1998: 126). The Fibonacci series has various interesting properties. It was found that when each successive number (after 3) is divided by its predecessor, the result approximated to 1.618. For example, 5 divided by 3 is 1.666 and 8 divided by 5 is 1.60. The higher up in the sequence, the closer the two successive numbers (one divided by the other) will approach the golden-section value of 1.6180 (or more precisely 1.6180339887). As noted previously, this golden-section value of 1.618 has become known as phi (Φ) and is closely associated with the construction known as the golden section.

Charles Bonnet (1720–1793 CE) observed that the spirals formed by the growth of leaves on a stem (known as phyllotaxis) were oriented clockwise and anticlockwise, invariably with frequencies conforming to successive numbers in the Fibonacci series. Edouard Lucas (1842–1891 CE) gave the numerical sequence first associated

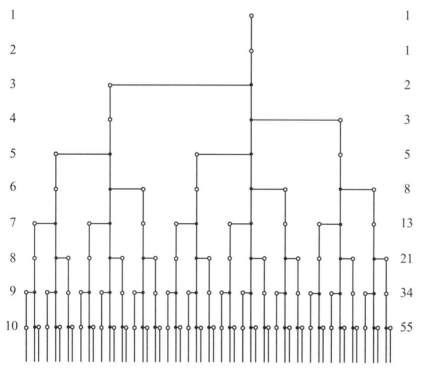

1

2

3

4

5

6

7

8

9

10

1

1

2

3

5

8

13

21

34

55

Figure 6.1 Rabbit family tree (JSS).

with Fibonacci its name: the Fibonacci series. Huntley (1970: 159–60) showed that the Fibonacci series was apparent in various other contexts, including the genealogical table of a particular drone or male bee. The drone hatches from a nonfertilized egg; fertilized eggs produce only females (queens and workers). So it can be calculated that a drone bee has one parent (a female), two grandparents (one male and one female), three great-grandparents (two female and one male) five great-great-grandparents (three females and two males) eight great-great-great-grandparents, and so on, thus yielding successive Fibonacci numbers (1, 2, 3, 5, 8, 13 . . .) at each stage of ancestry.

In the numerous publications and Internet Web sites dealing with the golden section and the Fibonacci series, attention is turned invariably to the incidence of Fibonacci series numbers in nature.

Often, lists of flowers with numbers of petals are presented. Certain contradictions can be noted. Thus Hemenway (2005: 136) lists lilies with 3 petals, buttercups with 5 petals, delphiniums with 8 petals, corn marigolds with 13 petals, asters with 21 petals and daisies with '34, 55 or 89 petals'. Huntley, on the other hand, claimed that a daisy may have 33 or 56 petals 'which just miss the Fibonacci numbers 34 and 55' (Huntley 1970: 161). Across the published material there appears to be occasional carelessness (some would term it selective reporting), with numbers selected simply because they are held within the numerical series and not because they are a true record of reality. For example, Hemenway (2005), by way of illustrating the persistence of Fibonacci numbers in petal counts, presents six unlabelled photographs of various flowers, including a photograph showing five full flower heads (a hybrid daisy of some

Plates 1 and 2 'Shiraz 1' (left) and 'Shiraz 2' (right), by Marjan Vazirian, were developed from façade tiling designs photographed in her home city of Shiraz (Iran).

Plate 3 Created by Craig S. Kaplan, this image shows an Islamic star pattern constructed from interlocking hexagonal rings embedded on the surface of a sphere. The pattern has been thickened into a delicate three-dimensional model and printed as a plastic sculpture using a rapid prototyping system.

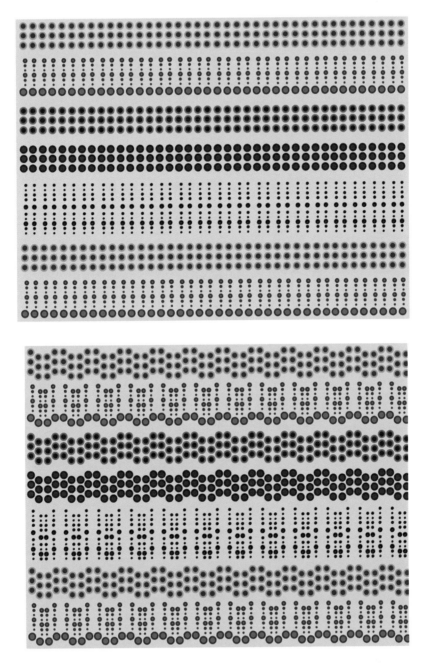

Plates 4 and 5 'Coronach 008' (top) and 'Coronach 013' (bottom), by the artist and academic Kevin Laycock, are visual responses following analysis of musical compositions by English composer Michael Berkeley.

Plate 6 'Three Fishes', by Robert Fathauer, is a limited-edition screen print made in 1994. The motif bears a resemblance to one of M. C. Escher's tessellation designs, but the symmetry of the design is quite different. Escher's design has four-fold rotational symmetry about the tail and the top fin, with two-fold rotational symmetry about the jaw. 'Three Fishes', on the other hand, has three-fold rotational symmetry about the tail, top fin, and mouth.

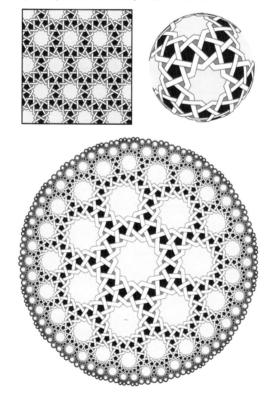

Plate 7 Craig S. Kaplan is based at the University of Waterloo in Canada. His research focuses on the relationships between computer graphics, art and design, with an emphasis on applications to graphic design, illustration and architecture. This illustrates a traditional Islamic star pattern, drawn with interlaced bands. The pattern is shown in its usual form in the upper left. The upper right and bottom images show re-imaginings of the same pattern on the surface of a sphere and in the curved geometry of the non-Euclidean hyperbolic plane.

Plate 8 'Point and Line', textiles from the University of Leeds International Textiles Archive (ULITA).

Plate 9 'Point and Line', textiles from the University of Leeds International Textiles Archive (ULITA).

Plate 10 'Point and Line', textiles from the University of Leeds International Textiles Archive (ULITA).

Plate 11 'Point and Line', textiles from the University of Leeds International Textiles Archive (ULITA).

Plate 12 'Point and Line', textiles from the University of Leeds International Textiles Archive (ULITA).

Plate 13 'Point and Line', textiles from the University of Leeds International Textiles Archive (ULITA).

Plate 14 'Point and Line', textiles from the University of Leeds International Textiles Archive (ULITA).

Plate 15 'Point and Line', textiles from the University of Leeds International Textiles Archive (ULITA).

Plates 16 Designs by Guido Marchini. Courtesy of the University of Leeds International Textiles Archive (ULITA).

Plates 17 Designs by Guido Marchini. Courtesy of the University of Leeds International Textiles Archive (ULITA).

Plates 18 Designs by Guido Marchini. Courtesy of the University of Leeds International Textiles Archive (ULITA).

Plates 19 Designs by Guido Marchini. Courtesy of the University of Leeds International Textiles Archive (ULITA).

Figure 6.2a–e Non-Fibonacci number of petals, Leeds, 2011.

kind) with what appear to be 18, 17, 19, 17 and 14 petals—all numbers missing from the Fibonacci series (Hemenway 2005: 136). Although Hemenway (2005) does not argue that the persistence of Fibonacci numbers in flower-petal counts is a fixed rule, this is the implication across much of the other relevant literature, and there seems to be the desire to ignore the fact that non-Fibonacci numbers can also be detected readily in petal counts and in nature in general (see e.g. Figure 6.2a–e).

the golden section

As indicated in the introduction, the golden section is the term given to a line divided into two unequal segments, so that the shorter segment is to the longer segment as the longer segment is to the whole line (Figure 6.3). Assuming the longer segment is equal to one unit of measurement, then the whole line will be equal to 1.6180, and the shorter segment thus 0.6180. So it can be seen that two lengths are considered to conform to the golden ratio if the longer length is equal to 1.6180 times the smaller length. The golden-section ratio is therefore 1.6180:1 or Φ:1.

As noted by Kappraff (1991: 83), Paul Klee in his *Notebooks* (1961) presented a simple means of subdividing a line (using only a pair of compasses and straight edge) into golden-section segments. This means of construction is known as the 'triangle-construction method' (Elam 2001: 26). Starting with line AB (as shown in Figure 6.4), draw AC =

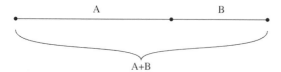

A+B is to A as A is to B

Figure 6.3 Golden section (AH).

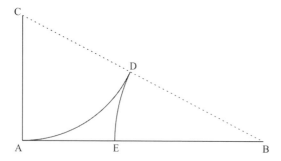

Figure 6.4 Golden-section triangle construction method (AH).

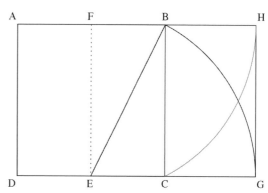

Figure 6.6 Construction of golden-section rectangle (AH).

If diagonals are drawn on pentagon ABCDE (Figure 6.5), points of intersection F and G divide each diagonal into golden-section segments.

the golden rectangle

A golden-section rectangle (sometimes referred to simply as a golden rectangle) can be constructed from a square as follows: starting with square ABCD (as in Figure 6.6), bisect side CD at E and side AB at F. Draw line EB (sometimes referred to as a diagonal to the half). Using E as centre, draw an arc with radius EB to line DC. Extend DC to intersect the arc at G. Draw an arc of radius FC using F as centre and extend line AB to intersect the arc at H. Complete the rectangle with sides in the ratio of 1.6180:1 by drawing line GH. This golden-section rectangle (AHGD) thus consists of an initiating square (ABCD) and a smaller rectangle (BHGC). This smaller rectangle (sometimes called the reciprocal golden rectangle), like its larger parent, has sides in the ratio of 1.6180:1. Because of this particular characteristic, the golden-section rectangle is also known as the rectangle of the whirling squares (mentioned previously in Chapter 3, 'static and dynamic rectangles').

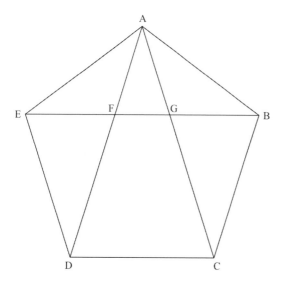

Figure 6.5 Golden-section segments created by diagonals in pentagon (AH).

1/2 AB perpendicular to AB. Draw arc CA to intersect CB at D, and arc BD to intersect AB at E, thus dividing AB into golden-section segments BE and AE at a ratio of approximately 1.6180:1.

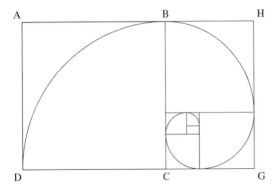

Figure 6.7 Golden-section spiral (AH).

the golden spiral

Following from the golden-section-rectangle construction shown above, further subdivisions can be applied and further reciprocal rectangles created. A golden-section spiral (also simply known as the golden spiral) can be created from this construction by using the sides of the square contained in successive reciprocal rectangles as successive radii of arcs connected to each other (Figure 6.7).

uses in art, design and architecture—a summary of the debate

There is a rich seam of studies which argue for the detection of the golden section and related measures in architecture, painting and sculpture as well as in design. Ghyka (1946), Doczi (1981), Lawlor (1982), Elam (2001) and Hemenway (2005) represent an informative cross-section of relevant literature, readily accessible to most art-and-design students and probably stocked in most art-and-design libraries, at least within Europe and North America. Typically, a golden-section analysis would be conducted as follows:

a building or work of art is identified, a line drawing of a façade, sculpture or painting or other object is presented, a drawing of a golden-section rectangle (with sides in the ratio 1.6180 to 1) is superimposed on the line drawing and note is made of the position of key aesthetic or structural elements of the drawing and how these coincide with diagonals or other divisions within the golden-section rectangle. Typically, discussions which follow such analyses argue (or at least imply) the conscious application by artist, craftsperson or builder of golden-section proportions in the creation or manufacture of the selected work. Occasionally, there is the suggestion that the artist, although unaware of golden-ratio theory, is imitating nature's proportions. In the late-twentieth and early-twenty-first centuries, there was a trickle of scholarly articles (mainly by mathematicians) challenging the well-established views on the use of the golden section in art, design and architecture and the apparent lack of scholarly rigour on which such views have been based; the most notable critics are Fischler (1979, 1981a,1981b), Markowsky (1992), Ostwald (2000), March (2001), Falbo (2005) and Huylebrouck (2009).

It has been claimed that the concentric circles of Stonehenge (3100–2200 BCE) have golden-section proportions (Doczi 1981: 39–40). However, the evidence given for such an interpretation is rather sparse, and such a relationship between the circles/henges seems to be based largely on surmise and assumption. It may well be the case that certain important mathematical ratios were indeed used during the various stages of construction, but the evidence presented by Doczi is not convincing.

It has been conjectured often that the ancient Greeks employed the golden section consciously in architecture and sculpture. Numerous studies have claimed that the proportions of the Parthenon are evidence of a deliberate use of the golden section in Greek architecture. Often, in relevant texts the façade as well as elements of the façade are circumscribed by golden rectangles (with sides in the ratio of 1.6180:1). There are substantial variations or dissimilarities from study to study. Huylebrouck and Labarque (2002), for example, presented a strong case contradicting several studies which claimed the detection of golden-section rectangles in the east façade of the Parthenon. Gazalé (1999) observed that it was seemingly not until the time of Euclid's *Elements* (308 BCE) that the golden section's mathematical properties were considered. This was some time after the building of the Parthenon (447–432 BCE); importantly, Euclid treated the golden section as any other geometric measure and did not assign any special significance to it (Gazalé 1999). So it may well be the case that the perceived connection between the golden section and ancient Greek architecture and sculpture is without foundation and is not supported by actual measurements. Various articles by Fischler (1979, 1981a and 1981b) and Markowsky (1992) offer a strong challenge to claims that the golden section was commonly used by builders and artists before the beginning of the modern era. Despite the scholarly weight of these studies and the solidity of the arguments offered, the catalogue of publications claiming the detection of golden-section measures in past art, architecture and design continued to grow, at least until the first decade of the twenty-first century. The conscious use of the golden section is what is implied in much of the relevant literature, rather than the selection of particular dimensions, proportions or compositional measures based simply on artistic/aesthetic judgment.

Leonardo da Vinci's work is often the subject of attention among golden-section researchers

(Hemenway 2005: 112–13). Making reference to the *Mona Lisa,* Livio (2002) summarized the situation by observing that this particular work has been the subject of such large quantities of contradictory speculation from both scholars and other observers that it is 'virtually impossible' to reach an unambiguous conclusion with respect to the golden section (Livio 2002: 162). A similar sentiment could be expressed for many of the publications focused on detecting golden-section proportions in the work of Michelangelo, Dürer, Seurat, Rembrandt and Turner (e.g. Hemenway 2005). Gothic cathedrals and various Egyptian pyramids have received similar treatment with similarly unconvincing results.

the potential for the creative practitioner

Although convincing evidence for the use of the golden section and golden-section measurements in past art, design and architecture is sparse and past analyses are weak, the potential benefits of the systems proposed by certain observers should not be dismissed. Indeed various twentieth-century artists and designers deliberately applied the associated measures, proportions or ratios to their work. The work of Le Corbusier and Hambidge is of particular note.

Charles-Edouard Jeanneret (known as Le Corbusier), the eminent twentieth-century architect, developed a rule of design known as the modular, a system based on the use of two numerical series (one known as the red series and the other as the blue series). In the book entitled *The Modular,* Le Corbusier explained that his system was based on the human figure (a six-foot-tall male) and also on a series of related golden sections and Fibonacci numbers (1954: 55). An important point to note is that Le Corbusier's intention was not to propose a restrictive set of rules to be strictly adhered to, but rather to provide a flexible framework which could be bent, manipulated or adjusted, particularly in circumstances where strict adherence goes contrary to the designer's intuitive judgment (Le Corbusier 1954: 63).

An important aspect of Le Corbusier's modular design system is that it is based on human proportions, a design outlook which can be traced back to the time of Vitruvius, the Roman engineer

and architect, as well as to various Italian Renaissance thinkers such as Luca Pacioli (whose treatise *Divina Proportione* was illustrated by Leonardo da Vinci). A series of rectangles in ratios relating to his two numerical series was also presented by Le Corbusier.

Probably the most interesting and potentially useful proposal from the viewpoint of the modern artist and designer is that put forward by Hambidge, a few decades before Le Corbusier. Hambidge also presented a catalogue of rectangles based largely on root rectangles (dealt with previously in Chapter 3) as well as whirling-square rectangles (created within golden-section rectangles). He referred to both these categories as designer's rectangles.

A compositional rectangle with sides in the ratio of 1.618:1 is presented here and divided in the same way as the root rectangles were divided

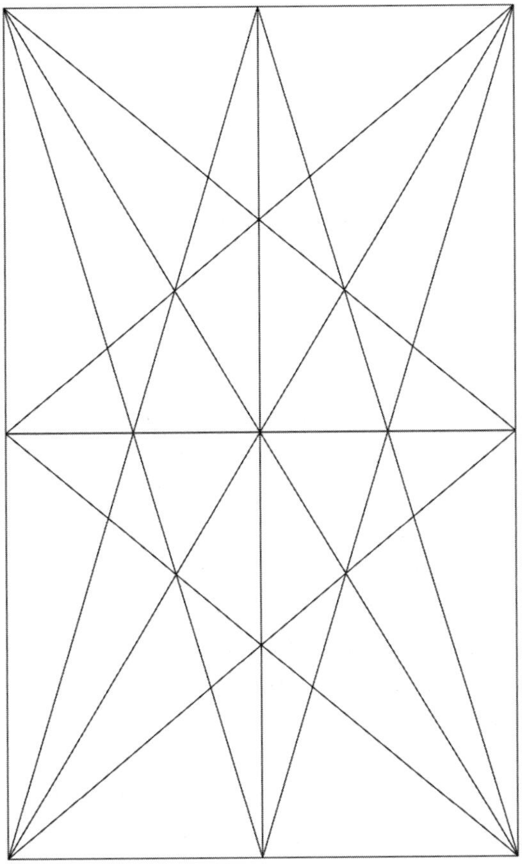

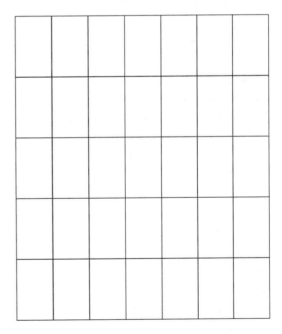

Figure 6.8 Brunes-Star-type golden-section rectangle (AH).

Figure 6.9 Rectangular grid with cells in approximate ratio of golden section (AH).

in Chapter 3. So, two corner-to-corner diagonals are drawn and the centre of the rectangle is thus identified. Each angle at the centre is bisected, and lines are drawn to connect the mid-points of opposite sides, thus allowing the rectangle to be divided into four equal parts. Next, the eight half diagonals (two diagonals from each angle to the mid-points of the two opposite sides) are drawn. A Brunes-Star-type division has been created, though not based on a square. Again, where lines overlap, key aesthetic points (KAPs) are identified (Figure 6.8).

Also presented is another grid with a golden-section-rectangle cell (in the ratio of 1.618:1) (Figure 6.9). This can be considered to supplement the various compositional aids given elsewhere in the book. The reader is invited to use these as frameworks on which to plan his/her own art-and-design work. Like Le Corbusier, the artist or designer should 'reserve the right at any time to doubt the solutions furnished…by such a system and should follow individual aesthetic judgment rather than adhere to a plan which runs contrary to such instinct' (Le Corbusier 1954: 63).

CHAPTER SUMMARY

This chapter has reviewed briefly the nature of the golden section and related geometrical figures such as the golden-section rectangle (also known as the whirling-squares rectangle, mentioned previously in Chapter 3) and the golden-section spiral. The Fibonacci series was explained and particular attention was turned to the use of the golden section and related figures or measures in art, design and architecture. A compositional rectangle and grid were proposed for possible use by today's creative practitioners.

With some exceptions, research aimed at finding the golden section and related measures in past art or design work is burdened with weak methodology, superficial analysis and largely unhelpful commentary, all under the guise of academic research. Methods used in analysis should be clear and unambiguous in order to ensure that data obtained are reproducible from researcher to researcher. Replication of results is an important issue, for theory building and progress cannot occur without it.

polyhedra, spheres and domes

introduction

The bulk of attention so far in this book has been on two-dimensional phenomena such as grids, tilings, patterns and ways in which a designer may plan or use two-dimensional space. Two important categories of tilings, known as the Platonic (or regular) tilings and the Archimedean (or semi-regular) tilings, were identified in Chapter 4. In three-dimensional space, certain arrangements of regular polygons can also be grouped into two classes, again using the names of the same two Greek thinkers. The first class is referred to as the Platonic, or regular, polyhedra or solids, and the second class is referred to as the Archimedean, or semi-regular, polyhedra or solids. As with tiling arrangements there are several other classes of polyhedra, but it is only the two classes mentioned here which are given attention in this chapter. Brief consideration is given also to the nature of a sphere and to the presence of polyhedra in various contexts, including their use in art, design and architecture.

spheres

In many textbooks aimed at characterizing the nature of three-dimensional figures or objects, early reference is made to the sphere. A sphere can be considered to relate to three dimensions as a circle relates to two dimensions, and can be visualized best as resulting from the rotation in three-dimensional space of a semicircle about its diameter (Ching 1996: 42). Often, the Platonic and Archimedean solids (covered briefly in sections below) are considered in relation to the sphere, and are occasionally represented inscribed within one. The sphere is perfectly circular (or round) with regular or even curvature across its surface, and infinite symmetry in terms of both rotation and reflection about its centre. All points on the surface are equidistant from the centre and form radii with the centre. A line which takes a path of maximum distance through the sphere is a diameter (twice the length of the radius). There are, in principle, an infinite number of radii and diameters. Each sphere may be divided into two hemispheres by simply passing a flat plane through its centre. Various rules of geometry describing two-dimensional space are not applicable to spheres and other curved objects; for example, the sum of the interior angles of a triangle created on the surface of a sphere exceeds 180 degrees. When compared with all other known solid objects, the sphere can hold the greatest volume per unit of its surface area.

Platonic solids

In the context of this book, a polyhedron is best considered to be a solid three-dimensional object

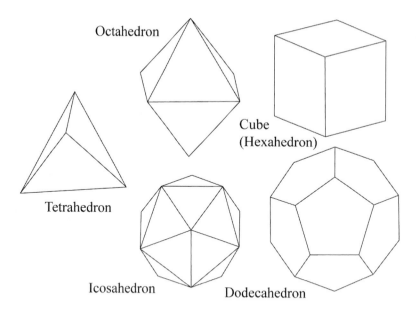

Figure 7.1 The five Platonic solids (MV).

Octahedron

Cube
(Hexahedron)

Tetrahedron

Icosahedron   Dodecahedron

made up entirely of polygonal faces, two joined at each straight edge and three or more meeting at corners (or vertices) in such a way that there is no gap or overlap. Sometimes a polyhedron is considered to include only its exterior surface, sometimes the volume held within its faces (in cases where each face is a closed plane) is included and sometimes it is considered in skeletal form and thus consisting of edges and vertices only. There is a total of five Platonic solids (Figure 7.1); these are polyhedra which can be constructed under the proviso that in each case all faces are identical regular polygons, and that the same number of component polygons meets at each vertex. The Greeks are credited with the identification of these five solids and provide the earliest written evidence (e.g. Plato's *Timaeus*). Further explanation is given in the following paragraphs.

On a given Platonic solid, the faces are thus identical in size and shape, and, as stated before, an equal number of faces meets at each vertex (so all vertices are equal). A pyramid, with a square base and triangular sides, is thus excluded. Three of the Platonic solids have equilateral-triangle faces, with three, four or five meeting at each vertex; each of these polyhedra has a name which makes reference to the constituent number of faces: tetrahedron (four faces), octahedron (eight faces) and icosahedron (twenty faces). The two additional polyhedra in this series are the cube (or hexahedron) with six square faces, and the dodecahedron with twelve regular-pentagon faces. Each Platonic solid can be considered held within a sphere with each vertex touching the inner surface of the sphere. It should be noted that, although the topic is not dealt with in this book, polyhedra, like patterns and tilings except in three dimensions, exhibit symmetry characteristics of rotation and reflection, with rotation axes and reflection planes operating through the body of the relevant solid. The symmetry characteristics of the Platonic solids were dealt with by Thomas and Hann (2007) and formed an important component part of Thomas's PhD thesis, which was focused on solving the problems encountered when tiling these three-dimensional solids with forms of decoration without leaving gaps or making overlaps (Thomas and Hann 2008).

As noted previously, each regular polyhedron has identical regular faces and identical vertices. It can be seen that there are only five possibilities when the nature of combining polygons in three-dimensional space is considered. At least three polygons are needed to form a solid polyhedral angle (i.e. an angle in a three-dimensional figure). This is possible with three, four and five equilateral triangles around a point (Figure 7.2a–c). With

six equilateral triangles the assembly lies flat (Figure 7.2d). Three squares make a solid angle, but four squares lie flat (Figure 7.3a–b). Three regular pentagons can form a solid angle but when lying flat form a gap insufficient for another equal pentagon (Figure 7.4). Three hexagons together lie flat and form the basis of one of the Platonic tilings, but are unable to make a solid polyhedral angle (Figure 7.5). Higher-order polygons cannot be used on their own to create a solid polyhedral angle around a vertex point, so a limit has been reached. Therefore, only three regular polygons are capable on their own of forming solid

Table 7.1 PROPERTIES OF PLATONIC SOLIDS

Solid	Faces	Edges of Face	Vertices	Edges at Vertex	Edges
Tetrahedron	4	3	4	3	6
Cube	6	4	8	3	12
Octahedron	8	3	6	4	12
Dodecahedron	12	5	20	3	30
Icosahedron	20	3	12	5	30

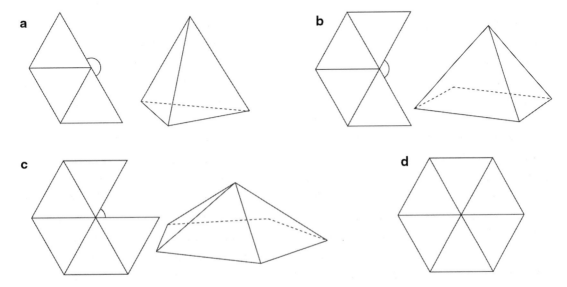

Figure 7.2a–d Three, four, five and six equilateral triangles (MV).

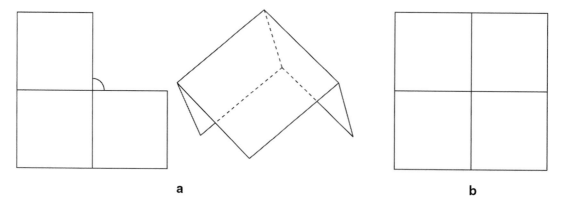

a   **b**

Figure 7.3a–b Three and four squares (MV).

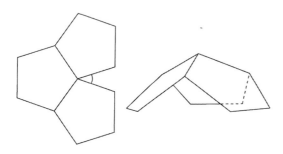

Figure 7.4 Three regular pentagons (MV).

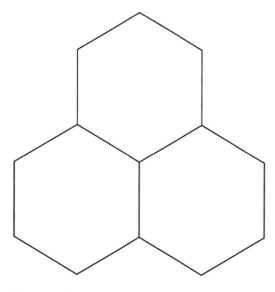

Figure 7.5 Three regular hexagons (MV).

polyhedral angles at vertices: the equilateral triangle (tetrahedron, octahedron and icosahedron), the square (hexahedron or cube) and the pentagon (dodecahedron). Proof of only five regular polyhedra was given by Euclid, the ancient Greek geometer (Heath 1956, vol. 3: 507–8). The properties of the five Platonic solids are summarized in Table 7.1. Each of the five is considered briefly in the next paragraphs.

The simplest of the five regular polyhedra is the tetrahedron, with six edges and four equilateral-triangle faces, three meeting at each of its four vertices (Figure 7.6). In ancient times, the tetrahedron was associated with the element of fire. Throughout the literature, flattened forms or nets of each polyhedron are illustrated, showing each of the faces connected to an adjacent face, as if like a children's cut-out-and-glue-the-flaps-together toy. An interesting paper titled 'Pull-up Patterned Polyhedra: Platonic Solids for the Classroom' was written by Meenan and Thomas (2008). The relevant net for the tetrahedron is shown in Figure 7.7.

The octahedron has twelve edges and eight equilateral-triangle faces, with four meeting at each of the six vertices (Figure 7.8). In ancient times, the octahedron was associated with the element of air. The relevant net is shown in Figure 7.9.

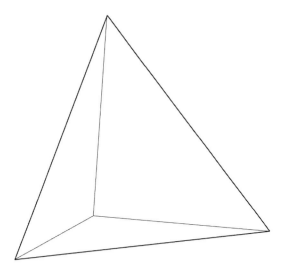

Figure 7.6 Tetrahedron (JSS).

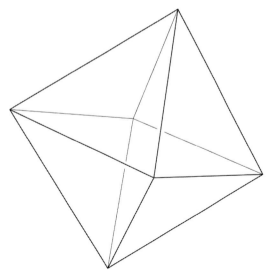

Figure 7.8 Octahedron (JSS).

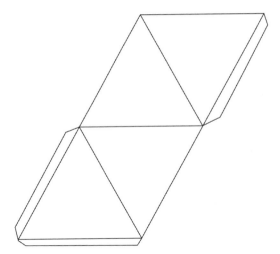

Figure 7.7 Tetrahedron net (JSS).

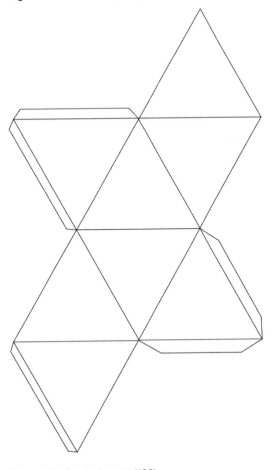

The icosahedron has thirty edges and twenty equilateral-triangle faces, with five meeting at each vertex (Figure 7.10). The icosahedron was associated with water in ancient times. The relevant net is shown in Figure 7.11.

The cube has twelve edges and six square faces, with three meeting at each of the eight vertices (Figure 7.12). In ancient times, the cube was associated with earth. A relevant net (there

Figure 7.9 Octahedron net (JSS).

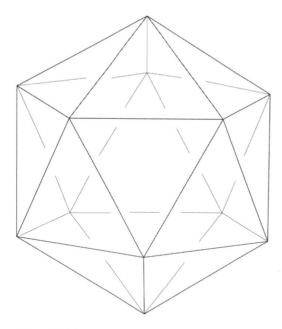

Figure 7.10 Icosahedron (MV).

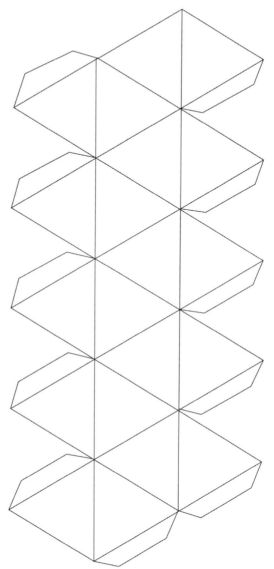

Figure 7.11 Icosahedron net (JSS).

are eleven possibilities) for the cube is shown in Figure 7.13.

The dodecahedron has thirty edges and twelve regular pentagonal faces, with three meeting at each of the twenty vertices (Figure 7.14). In ancient times, the dodecahedron was associated with the universe/cosmos. A net for the dodecahedron is given in Figure 7.15.

A further important characteristic of the set of Platonic solids is how they relate to each other. It has been stated previously (and listed in Table 7.1) that the cube has six faces and eight vertices, and the octahedron has six vertices and eight faces. This relationship, known as duality, can be ascertained as follows. Consider the cube. When the middle of each face is marked with a point and the shortest distance between adjacent points is connected by straight lines, the edges of the octahedron are drawn, with vertices forming at the centre of faces, and the octahedron itself forming within the cube. Likewise, when the same procedure is followed for the octahedron, a cube is outlined. The cube and octahedron are thus considered to be duals to each other. Meanwhile, the dodecahedron, with twelve faces and twenty vertices, and the icosahedron, with twelve vertices and twenty

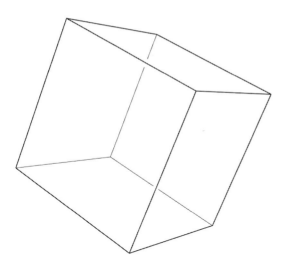

Figure 7.12 Cube (JSS).

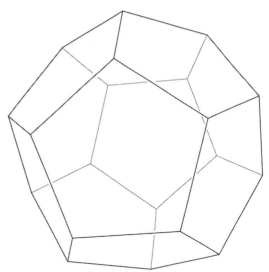

Figure 7.14 Dodecahedron (JSS).

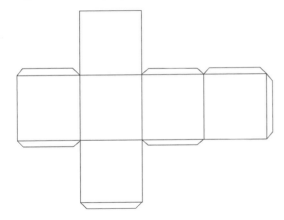

Figure 7.13 Cube net (JSS).

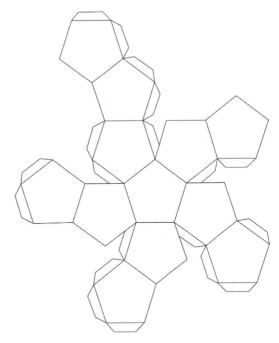

Figure 7.15 Dodecahedron net (JSS).

faces, are duals to each other in the same way. The remaining Platonic solid, the tetrahedron, with four faces and four vertices, is considered to be a dual to itself: when centres of faces are connected, an inverted tetrahedron is formed within a tetrahedron.

Archimedean solids

The Platonic solids form the basis of another set of polyhedra. Through a process known as truncation (slicing off edges and/or vertices) it is possible to create a set of thirteen polyhedra known as either the Archimedean solids or semi-regular polyhedra. These thirteen solid forms consist of regular polygonal faces of more than one type, with a specific number and type of polygon meeting at each vertex of each solid. In each case, vertices are thus identical, as with the Platonic solids. Each of the thirteen can fit perfectly within a sphere. Six of the Archimedean solids are derived (or truncated) from the cube and octahedron, and six others from the icosahedron and dodecahedron. The thirteenth is obtained through truncating the tetrahedron (Kappraff 1991: 329). The full set is identified and illustrated in Figure 7.16. Another process, known as stellation, simply adds further structures (such as pyramid shapes) to the faces of each of the five Platonic solids, giving many of them a star-type characteristic. Holden (1991) provided a review covering a wide range of three-dimensional objects and showed the potential variation attainable through truncation and stellation.

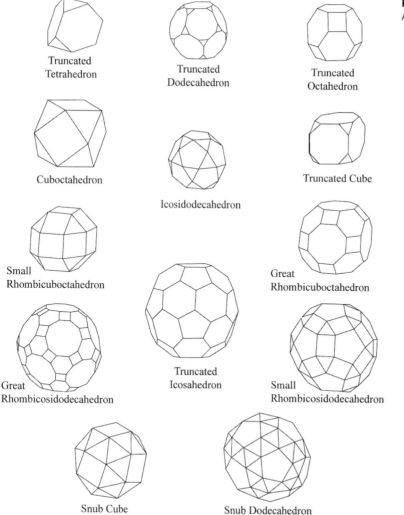

Figure 7.16 The thirteen Archimedean solids (MV).

Truncated Tetrahedron

Truncated Dodecahedron

Truncated Octahedron

Cuboctahedron

Icosidodecahedron

Truncated Cube

Small Rhombicuboctahedron

Great Rhombicuboctahedron

Great Rhombicosidodecahedron

Truncated Icosahedron

Small Rhombicosidodecahedron

Snub Cube

Snub Dodecahedron

radiolarians, soccer balls and super molecules

Radiolarians (also radiolaria) are tiny sea creatures (known as holoplanktonic protozoa) distributed widely in the oceans of the world; the skeletons of these creatures yield an amazing diversity of forms. *Kunstformen der Natur* (1904) by Ernst Haeckel (1834–1919) included numerous lithographs depicting radiolarians and other sea creatures. This publication influenced early-twentieth-century art, design and architecture to a great degree. It is worth noting in particular that several of the illustrations (produced by lithographer Adolf Giltsch from sketches made by Haeckel) show close structural similarities to several of the polyhedra mentioned above.

One particular polyhedron, the truncated icosahedron (one of the Archimedean solids listed in Table 7.2), is worth identifying as it seems to appear in a range of guises (Figure 7.17). The polyhedron consists of twelve regular pentagonal faces and twenty hexagonal faces, with sixty vertices and ninety edges; each vertex is the meeting-point of two hexagons and one pentagon. The structure is easily recognizable and was used in the form of a soccer ball in the late-twentieth century. Interestingly, Leonardo da Vinci produced an illustration of the truncated icosahedron for Luca Pacioli (mentioned previously and in the next section) in the fifteenth century. The structure is also found in a particular molecular form, known as C60 Buckminsterfullerene (nicknamed Buckyball) after the eminent engineer and architect R. Buckminster Fuller due to its close resemblance to various dome structures (known as geodesic domes or spheres) which had been designed by Fuller.

polyhedra in art and design

Quantities of carved stones showing the symmetries (or underlying geometry) of all five regular polyhedra have been found in Scotland and may be as much as 4,000 years old; these are on display in the Ashmolean Museum in Oxford. Polyhedra have been associated with art and design for many centuries. A notable peak was reached during the Renaissance in Europe. The well-known Renaissance artist Piero della Francesca (ca. 1415–1492) was an accomplished geometer and authored several books on the subject of geometry (including the regular solids) and the then-developing field of linear perspective. These works were to influence later mathematicians and artists, including Luca Pacioli and Leonardo da Vinci. In his *Lives of the Painters,* Vasari remarked that Piero della Francesca's 'writings on geometry and perspective . . . are inferior to none of his time' (quoted in Seeley 1957: 107).

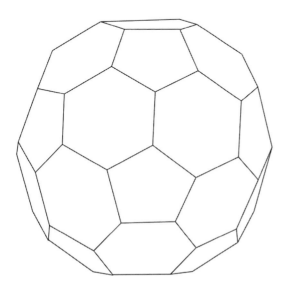

Figure 7.17 Truncated icosahedron (MV).

Table 7.2 ARCHIMEDEAN SOLIDS (OR SEMI-REGULAR POLYHEDRA)

Solid	F	V	E	CP	VA
Truncated Tetrahedron	8	12	18	4 × 3-sided 4 × 6-sided	3.6.6
Truncated Dodecahedron	32	60	90	20 × 3-sided 12 × 10-sided	3.10.10
Truncated Octahedron	14	24	36	6 × 4-sided 8 × 6-sided	4.6.6
Cuboctahedron	14	12	24	8 × 3-sided 6 × 4-sided	3.4.3.4
Icosidodecahedron	32	30	60	20 × 3-sided 12 × 5-sided	3.5.3.5
Truncated Cube	14	24	36	8 × 3-sided 6 × 8-sided	3.8.8
Small Rhombicuboctahedron	26	24	48	8 × 3-sided 18 × 4-sided	3.4.4.4
Great Rhombicuboctahedron	26	48	72	2 × 4-sided 8 × 6-sided 6 × 8-sided	4.6.8
Great Rhombicosidodecahedron	62	120	180	30 × 4-sided 20 × 6-sided 12 × 10-sided	4.6.10
Truncated Icosahedron	32	60	90	12 × 5-sided 20 × 6-sided	5.6.6
Small Rhombicosidodecahedron	62	60	120	20 × 3-sided 30 × 4-sided 12 × 5-sided	3.4.5.4
Snub Cube	38	24	60	32 × 3-sided 6 × 4-sided	3.3.3.3.4
Snub Dodecahedron	92	60	150	80 × 3-sided 12 × 5-sided	3.3.3.3.5

Key: F = faces; V = vertices; E = edges; CP = constituent regular polygons; VA = vertex arrangement.
Source: Data taken from A. Holden (1991: 46) and B. G. Thomas (2012).

A marble floor mosaic in the Basilica di San Marco (Venice) features a small stellated dodecahedron attributed to Paolo Uccello (1397–1475). *De Divina Proportione* (1509) by Lucia Pacioli (1445–1517) dealt mainly with mathematical and artistic proportion. Leonardo da Vinci (1452–1519), who apparently took mathematics lessons from Pacioli, provided illustrations of polyhedra in skeletal form (see-through images, featuring edges and vertices only) using linear perspective;

this meant that the viewer could consider more fully the three-dimensional characteristics of these solid objects rather than simply imagining their nature from cross-sectional or plan views. Fra Giovanni da Verona created a collection of intarsia (mosaic-type images of inlaid wood) depicting polyhedra around 1520.

Albrecht Dürer (1471–1528), the German printmaker, who may well have been influenced by the works of Luca Pacioli and Piero della Francesca, contributed to the literature of polyhedra through the publication of *Underweysung der Messung* (Dürer 1525), which dealt with the Platonic solids as well as other topics such as tilings, linear perspective and geometry in architecture. Although a few inaccuracies have been noted in Dürer's book, it seems that he was the first to introduce the concept of nets and unfolding polyhedra to provide an object similar in principle to a modern flat pack. Dürer's renowned engraving *Melancholia I* (1514) depicts a polyhedron as a seemingly important component of the composition.

M. C. Escher (1898–1972), the renowned Dutch artist, showed an interest in polyhedra, and depicted Platonic solids and related stellated forms (star-like polyhedra with small pyramid-type additions on each face) in a small selection of his work. In Salvador Dali's painting *The Sacrament of the Last Supper* (1955) Christ and his twelve Apostles are depicted within a giant dodecahedron (one of the Platonic solids).

domes

A dome is a structural feature of ancient architectural lineage resembling the hollow upper half of a sphere. Conveniently explained as an arch rotated around its vertical axis, a dome has good structural strength and can span substantial open spaces without assistance from further interior support. Technically advanced dome structures were created beginning well over two thousand years ago, from the time of the Romans (e.g. the Pantheon), Sassanians (from third- to eighth-century Persia) and the Byzantines to the later Ottoman, Safavid and Mughal empires in the sixteenth to eighteenth centuries. There is, however, much earlier indirect evidence in the form of an Assyrian bas-relief from Nimrod (seventh-century BCE) which depicts what appears to be a dome structure. Domes are features of many Russian Orthodox churches, St Peter's in Rome, St Paul's in London and the Taj Mahal in Agra.

Dome constructions probably evolved from the consideration of spheres (or related constructions) and their subdivision into networks of polygonal faces. In view of the fact that triangular structures offer the best strength-to-weight properties compared to other polygonal structures, triangular faces are often predominant in dome constructions. The icosahedron, with exactly twenty equilateral faces (the numerical limit of equilateral triangles in the construction of a convex polyhedron), is often the basis of the subdivision of the sphere and thus of related dome constructions.

In terms of structure, there are several types of dome. The onion dome is bulbous and tapers to a point. Often larger in diameter than the tower (or other construction) it rests on, the onion dome usually has a height which exceeds its width. The umbrella dome is segmented by radial spokes from top to around the base. A saucer dome is low-pitched and shallow. Corbel domes are constructed by positioning each horizontal row of building units (e.g. stones, cut blocks or bricks) further inwards relatively to the previous row until the final row meets at the top.

Figure 7.18 Eden project, showing biomes, photo by Jeremy Hackney.

Figure 7.19 General exterior view of Victoria Quarter, Belfast, photo courtesy of BDP.

In the modern era the term *geodesic* was introduced by the architect R. Buckminster Fuller, who conducted numerous experiments with structures based largely on a network of circles on the surface of a sphere which intersect to form triangular sections (together shaped like an icosahedron) which in turn distribute the stress across the full structure. There are numerous examples of geodesic-dome structures worldwide. Probably the most famous from the twentieth century are the Climatron (a green house, built in 1960, at the Missouri Botanical Gardens) and the Americam Pavilion used for Expo 67 at the Montréal World Fair 1967. The Eden Project (Cornwall,

Figure 7.20 General interior view of dome, Victoria Quarter, Belfast, photo courtesy of BDP.

Figure 7.21 Victoria Quarter, Belfast. View of dome from below, with internal and external cleaning galleries visible. The internal gantry contains a support structure for fabric sun-shade which tracks the path of the sun to prevent overheating on the upper viewing gallery. Photo courtesy of BDP.

United Kingdom) has two adjoining dome structures used to house plants from the world's tropical and Mediterranean regions. The gigantic domes (known as biomes), consist of hundreds of hexagonal and pentagonal inflated plastic (thermoplastic ETFE) cushion-like cells supported on tubular-steel frames (Figure 7.18). In the late-twentieth and early-twenty-first centuries, dome structures played an important role in architectural developments in many cities worldwide. An interesting example is the Victoria Quarter, Belfast (Figures 7.19–7.21).

CHAPTER SUMMARY

This chapter has introduced two classes of three-dimensional objects known as polyhedra, each consisting of assemblies of regular-polygon-shaped faces. Each of the so-called Platonic (or regular) solids, of which there are five types, has faces of only one regular polygon type: equilateral triangles in three cases and squares or regular pentagons in the other two. The Archimedean (or semi-regular) solids, of which there are thirteen types, consist of regular polygonal faces of more than one type with a specific number and type meeting at each vertex of each solid. The uses of polyhedra in art, design and architecture as well as their presence in various other contexts have been reviewed briefly.

structure and form in three dimensions

introduction

The two-dimensional world, consisting of length and breadth, is largely an imaginary (or representational) world which acts as host to images including geometric, figurative, or other types of drawing or writing, as well as various forms of surface decoration, televised, photographic or computer-screen images, including representations of three-dimensional scenes or objects.

The real world is a three-dimensional world comprised of length, breadth and depth. Comprehension of objects within this three-dimensional world requires more than one view, as such objects appear changed when seen from different distances or angles and under different lighting conditions. A sphere, or up-turned cylinder or cone, may appear disc-shaped if viewed from afar, and the true form may be realized visually only when viewed close up. Understanding three-dimensional reality requires collating visual information from different angles and distances. Wong observed that a designer working in three-dimensions 'should be capable of visualising mentally the whole form and rotating it mentally in all directions as if he had it in his hands' (1977: 6–7). Within three decades of Wong's comment, this visualization was assisted greatly by developments in computing technology. By the early-twenty-first century, most designers had access to computing software which made it possible to view on screen

a designed object from any selected angle, as if the object were lying in three-dimensional space. Also, rapid prototyping offered the facility of creating mock-ups for hand-held consideration.

An understanding of the geometric characteristics of the various polyhedra introduced in Chapter 7 can prove of value to the three-dimensional designer. There are, however, several other categories of form which have in the past played a substantial role in three-dimensional design development. These include prisms, cylinders, cones and pyramids and their numerous derivations. These are considered briefly in this chapter. Attention is turned also to how various forms can undergo transformation, using processes referred to as distortion, cutting and re-assembly, addition and subtraction. Adding to the introduction given previously (in Chapter 3) of various two-dimensional grids or lattices based on dynamic rectangles, this chapter presents a series of three-dimensional lattices based on similar proportions.

elements of three-dimensional form

As observed by Wong, 'similar to two-dimensional design, three-dimensional design also aims at establishing visual harmony and order, or generating purposeful visual excitement, except that it is concerned with the three-dimensional world'(1977: 6). As with two-dimensional design, the designer working in three dimensions needs to bring together various elements. Point, line, shape, plane, colour, texture, form and volume are all of importance. In can be seen that awareness of the various fundamental elements of design considered previously (especially in Chapter 2) is thus of importance to all designers, including those concerned with the various three-dimensional-design disciplines (e.g. product, fashion, industrial and architectural design). In addition, a small number of fundamental issues relating to structure and form in the three-dimensional context should not be ignored; these are considered briefly in the next paragraphs.

Wong considered shape to be 'the outward appearance of a design and the main identification of its type. A three-dimensional form can be rendered on a flat surface by multiple two-dimensional shapes' (1977: 10–11). When most three-dimensional objects are rotated in space, different shapes are revealed at different stages of rotation. Shape is only one aspect of form. Form is the total appearance of a design and takes into account size, colour and texture.

Structure is the 'skeleton beneath the fabric of shape, colour and texture' (Wong 1977: 14). The external appearance of a form may be complex, but the structure beneath may be comparatively simple. Volume relates to three-dimensional objects as area relates to objects on the two-dimensional plane.

Wong identified three 'primary directions' or views of importance to designers working in three dimensions, namely length, breadth and depth, which require measurement in vertical (up and down), horizontal (left and right) and transverse (forwards and backwards) directions (1977: 7). Selecting an imaginary rectangular prism by way

of example, these three dimensions can be represented by three intersecting rods (Figure 8.1a) or three intersecting planes (Figure 8.1b). Doubling and sliding-along of these constituent planes can create a rectangular prism, with the vertical

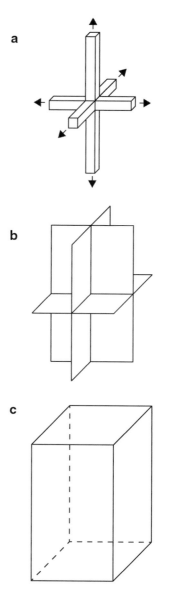

Figure 8.1a–c Three 'primary directions'; three intersecting planes; a prism formed from doubling three planes (JC, adapted from Wong 1977: 7).

directions represented by the front and rear panels (or faces) of the cube, horizontal directions by the top and bottom panels and transverse directions by the left and right panels (Figure 8.1c).

Wong noted the importance to the designer in three dimensions of what he called the 'relational elements': position, direction (or orientation) space and gravity (1977: 12). These bring together the structural elements which in turn determine the form of the design. Position of a point within a prism can be considered by reference to distance from the three pairs of faces of the prism (Figure 8.2a–b). Position is thus the location of an object relative to its surroundings. Direction or orientation of a line within the prism can be considered by reference to the angle of the line relative to the faces of the prism (Figure 8.2c–d). The space associated with a design can be considered as occupied if a solid object such as a die is considered, or hollowed if each face of the object was constructed from thin wire or straws. As noted by Wong gravity has an effect on the stability of an object, and the behaviour of all three-dimensional objects is influenced by the laws of gravity (1977: 12). The material contents of a form will determine its weight. Lead is heavy, and feathers are light. Unsupported, a cube, for example, will fall from mid-air; support or anchoring is required. With one of its faces resting on a firm surface, a cube is stable, but when attempts are made to rest it on one of its vertices, gravity forces it to topple. Form and weight, as well as the pull of gravity, are determinants of stability.

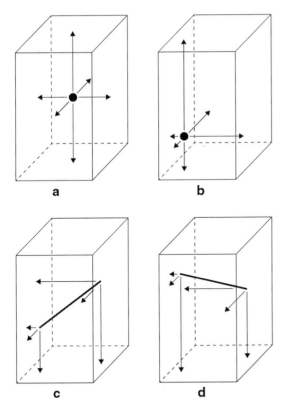

Figure 8.2a–d Position of a point can be considered by reference to the distance from faces of the prism (a and b). Direction of a line can be considered by reference to its orientation (or angle relative to faces) within the prism (c and d) (JC, adapted from Wong 1977: 12).

representations of three-dimensional form

Designers need also to represent their designs visually in order that their appropriateness for the intended end-use, as well the ease with which they can be manufactured or built, can be assessed by others (typically the client or manufacturer). Drawings on a two-dimensional plane (e.g. a computer screen or a sheet of paper) are the means of achieving this objective. As observed by Ching (1998: 2), drawings of one kind or another are the common medium to guide a design idea from conception to the fully resolved constructed object. So, throughout the design process, drawings show the stages of development through to realization and also communicate the final design to the client or manufacturer. Different design disciplines (e.g. product or industrial design, architectural design, fashion or textile design) have different requirements and require different approaches. In each case the intention is to communicate both visual, technical and production information to ensure that the needs of the client, manufacturer, and market are addressed. Thus, there is a need for unambiguous visual information, and this is a key difference between technical-design drawing and expressive or creative drawing associated with the fine arts.

Drawings may be created through the use of one of several systems, classified according to the resultant effect as well as the method of projection. These systems were considered in some detail by Ching (1998) in the treatise *Design Drawing*; although intended principally (but not exclusively) as a source for professional architects and architecture students, this publication is of immense value to designers working in all disciplines.

With conventional drawing methods, there are several means of projecting a three-dimensional form on to a two-dimensional surface such as a sheet of paper. An important point to bear in mind is that with all manual drawing systems, certain visual information is lost through such projection. By the first decade of the twenty-first century, however, this weakness had been lessened to a considerable extent due to substantial strides in computer-aided-design technology, with varieties of software offering the possibility of a 360-degree vision of three-dimensional objects on screen. Specialist (book) publications, focused on the specific drawing needs of various design disciplines, have been published also.

The simplest and most common means of representing three-dimensional designs is through multi-view drawings. Three-dimensional designs

can be represented through the use of a series of flat views. In order to do this, it is convenient for designers to imagine that their design is placed within a cube and to present a series of directional views as if held within that cube. The principal views are plan view (or top view), elevation and

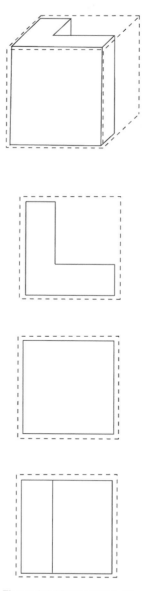

Figure 8.3 Views of three-dimensional figure placed within a cube (MV).

section view. Simple top, front and side views are shown in Figure 8.3. Many others are possible. For example, a total of six views can be realized simply by considering each face of the cube: a top plan view (as if from above); a bottom plan view (as if from below); a right-side view; a left-side view; a front view; a back view. Multi-view drawings provide representations of the external surfaces of the object. Supplementary or sectional views (as if the object has been cut into slices) may be useful also (Wong 1977: 16).

Three-dimensional scale models from wood, cardboard or other materials are a further means of representing a designer's intentions. Worthwhile guidelines were given by Wolchonok (1959) as well as Wong (1977). By the first decade of the twenty-first century, the demand for three-dimensional scale models had declined dramatically due to phenomenal strides in computing technology which, as indicated earlier, allowed three-dimensional designs to be represented realistically on screen and viewed from a multitude of angles. In addition, rapid prototyping techniques allowed precise scale models to be built from a range of materials. By the early years of the first decade of the twenty-first century, small-scale computer-aided manufacturing systems could be used to produce precise, functioning replicas of a three-dimensional design without the necessity for machine retooling and long production runs.

cubes, prisms and cylinders

When considering the fundamental structural aspects of three-dimensional forms, it is worth turning attention briefly to the geometry of the cube (one of the regular polyhedra dealt with in Chapter 7). This form is certainly of immense importance to all designers working in three dimensions. As recognized previously, the cube is the only regular polyhedron capable of space filling without gaps. Due largely to the equality of its dimensions, the cube is inherently stable when resting on a flat surface. However, it should be remembered that the strength-to-weight properties of a cube do not compare favourably with triangulated systems (Pearce 1990: xvii). Of course, other

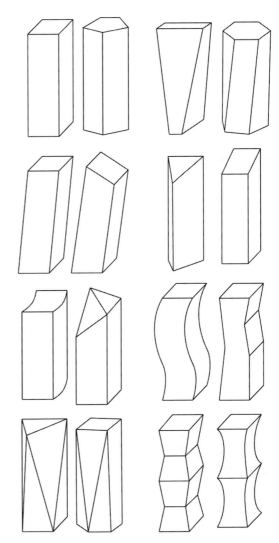

Figure 8.4 Conventional prism (top) and a selection of variations (JC, adapted from Wong 1977: 39).

structures can be cut from a cube. For example, six identical square pyramids can be cut, and if these are attached to the six faces of a second cube (with identical dimensions to the first cube), a rhombic dodecahedron is created. So the cube can act as a basis for the development of other three-dimensional forms.

Prisms are column-like forms with numerous cross-sectional possibilities, including square, triangular or other polygonal shapes. Ends may be flat, curved, pyramidal or irregular shapes. Edges may or may not be perpendicular to the ends, or may or may not be parallel to each other and ends may, for example, be square, triangular, hexagonal or pyramidal-shaped (examples shown in Figures 8.4). Wolchonok defined a 'prismatic surface' as 'a combination of plane surfaces in which the successive planes change direction' (1959: 4).

Most three-dimensional objects can be considered encased within a rectangular prism or some other standard three-dimensional form, and a process of removal or subtraction of parts of the prism can yield the form of the designed object. As observed by Wolchonok, in common practice the design of three-dimensional objects involves an 'additive process' as well as subtraction of parts (1959: 15). That is, components are added to the prism (or other standard three-dimensional form), and components are taken away. The challenge for designers is, of course, much greater than simply taking a solid shape and cutting away certain parts and adding others. Wolchonok commented: 'The designer's problem is not merely one of adding to nor subtracting from, but rather of working out a total object whose parts lend emphasis to the whole and whose overall aspect is an harmonious, integrated arrangement of related parts' (1959: 15–16).

Prisms are generally considered to be solid forms, although Wong presented a series of hollowed prisms and showed how these could be adjusted in various ways (mainly through cutting) to produce a wide range of forms of potential interest to designers working in three dimensions (1977: 40–2).

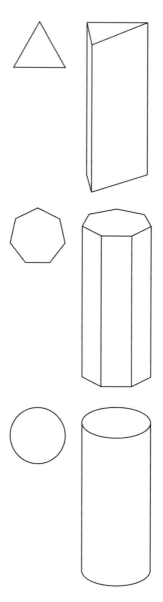

Figure 8.5 From prism to cylinder (JC, adapted from Wong 1977: 44).

A prism can be cut into smaller units or segments. Wolchonok presented a range of illustrative material showing prisms subdivided and cut in various ways (1959: 32–3). Of potential interest and value to the designer is the cutting of a prism into component parts and the rearrangement of those parts into forms which have a volume equal to that of the original prism. The prism can be cut into component parts, preferably making reference to one or more of the systems of proportion outlined in this book, and the component parts rearranged into an alternative form. Also, it may be of value to the designer to consider working from simple rectangular prisms with faces consisting of dynamic or golden rectangles.

Assuming a prism constructed from a series of flat, equal-shaped and equal-sized planes, the minimum number of planes required is three; this would result in a (hollow) prism with a triangular top, bottom and cross-section. As the number of planes (or faces) increases, with square, pentagonal, hexagonal and higher-order regular polygons used, further cross-sectional possibilities can be realized. Ultimately, by increasing the number of faces towards infinity, a circular cross-section

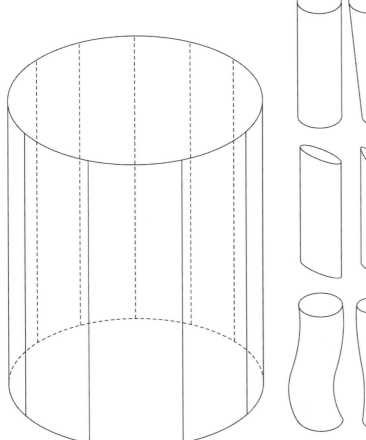

Figure 8.6 Conventional cylinder (AH).

Figure 8.7 Cylinder variations (JC, adapted from Wong 1977: 45).

develops, and another form, known as a cylinder is realized (Figure 8.5). Wong commented, 'The body of a cylinder is defined by one continuous plane, without beginning or end, and the top or bottom of a cylinder is in the shape of a circle' (1977: 44). Often, the cross-sections and ends of a cylinder are parallel circles of equal size (Figure 8.6) but, as shown by Wong (1977, p. 45), several deviations from this convention are possible. Examples of variations are shown also in Figure 8.7.

The cylinder is one of the most common curvilinear geometric shapes, conveniently envisaged as the circular motion of a line (hanging in three-dimensional space) about a straight-line axis with each point of the moving line tracing a circle. Alternatively, the same class of solid can be generated by the rotation of a rectangle (again hanging in three-dimensional space) about one of its sides (Ching 1996: 42). The cylinder is stable resting on a flat surface, provided it rests on its curved face.

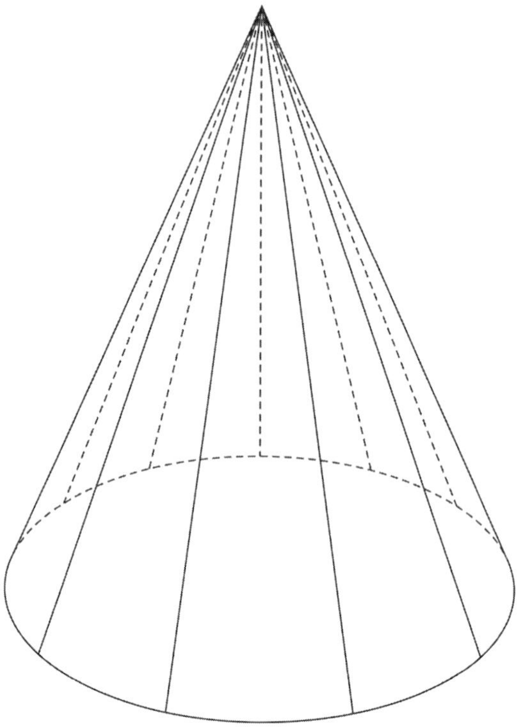

Figure 8.8 A cone (AH).

cones and pyramids

A cone (Figure 8.8) can be visualized as the resultant path of a straight line with one end held at a fixed point (the apex), and the other end tracing the circumference of a circle (or a related curved figure such as an oval shape). Alternatively, the object can be visualized as generated by the rotation of a right-angled triangle held in space, with one side acting as a stationary axis. The cone is stable on a flat surface when resting on its circular base.

A pyramid is a solid with a polygonal base and triangular faces meeting at a common point (often refereed to as an apex). The pyramid is stable on a flat surface if resting on one of its planes (either the base or one of its faces). It was noted previously that triangular planes were

used in the construction of three of the five Platonic solids: the tetrahedron, the octahedron, and the icosahedron. Triangular planes are also used in pyramid-shaped projections on various polyhedra. As noted in Chapter 7, 'domes', because of the particular strength-to-weight advantages offered by triangular planes, have achieved a position of considerable importance in three-dimensional design, particularly in dome-type architecture. Comparing the structural properties of a cube to those of triangular constructions, Pearce observed that 'There can be no doubt that cube oriented geometry is extremely important and relevant...however, it has serious modular limitations...as a structural framework it has inferior strength-to-weight properties when compared to triangulated systems' (1990: xvii).

three-dimensional lattices

In Chapter 5 it was seen that five two-dimensional frameworks or lattice structures, with accompanying unit cells acted as the basis for the seventeen symmetry classes of two-dimensional all-over patterns. Three-dimensional lattice structures are possible also and these have been employed by crystallographers as a means of considering the structure of three-dimensional forms at the microscopic level. Pearce (1990), in the treatise *Structure in Nature is a Strategy for Design,* considered these and related concepts and showed their usefulness in the development of novel three-dimensional architectural structures. Following from this, various three-dimensional lattice structures with unit cells based on a cube as well as the various dynamic rectangles are presented here. It should be remembered that lattice structures do not actually exist as collections of solid forms, but rather are a series of imaginary points associated with the vertices of a collection of imaginary unit cells. These series of imaginary points can act as compositional and structural guides in the design of three-dimensional constructions. An excellent, straightforward introduction to three-dimensional lattices and polyhedra was provided by Rooney and Holroyd (1994).

A cubic lattice, as the name implies, has cubic unit cells, each consisting of six equal-sized square faces. Eight unit-cell vertices meet at each lattice point (Figure 8.9). Further lattices can be proposed using unit cells comprised of golden rectangles (the ratio of 1:1.618) and root rectangles (referred to by Hambidge [1967] as dynamic rectangles). As noted previously, these have the following ratios: 1:1.4142 (the ratio of the root-two rectangle); 1:1.732 (the ratio of the root-three rectangle); 1:2 (the ratio of the root-four rectangle); 1:2.236 (the ratio of the root-five rectangle). Lattices based on each of the above are presented in Figures 8.10–8.14.

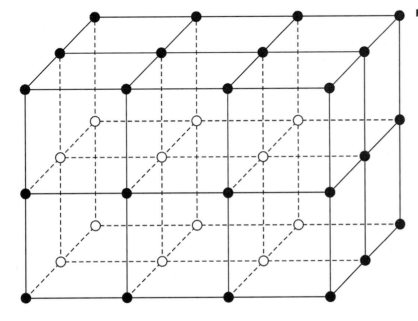

Figure 8.9 Square lattice (AH).

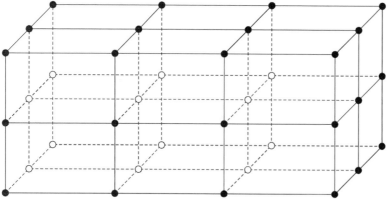

Figure 8.10 1 to 1.1618 lattice (AH).

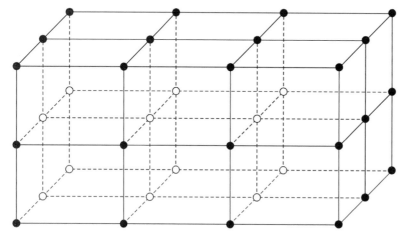

Figure 8.11 1 to 1.4142 lattice (AH).

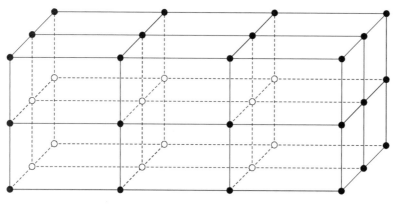

Figure 8.12 1 to 1.732 lattice (AH).

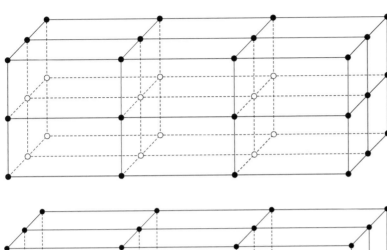

Figure 8.13 1 to 2 lattice (AH).

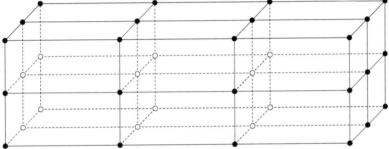

Figure 8.14 1 to 2.236 lattice (AH).

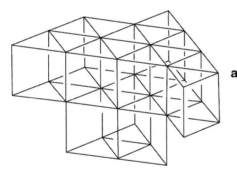

a

A triangular lattice structure consisting of triangular prism unit cells, each with two parallel triangular bases and three squares as faces (Figure 8.15a) can be proposed. Twelve unit-cell vertices meet at each lattice point. Similarly, a hexagonal lattice structure composed of hexagonal-prism unit cells, each with parallel hexagonal bases and six square faces, with all sides of equal length (Figure 8.15b) can be proposed. Six unit-cell vertices meet at each lattice point.

Each of the three-dimensional lattices or lattice structures suggested offers a safe guideline or framework for the disposition of components of a design in three dimensions. Numerous other possibilities can unfold, and the reader is encouraged to experiment.

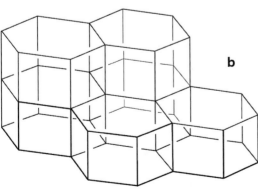

b

Figure 8.15 a–b Triangular and hexagonal three-dimensional structures (MV).

transformations

All the solid forms introduced in this book can be transformed through a range of techniques including distortion, cutting and re-assembly, addition and subtraction. Each is explained briefly here.

Distortion of a three-dimensional form is achieved through lengthening and/or shortening of one or more dimensions. A sphere, for example, can be transformed by elongation along a diameter to provide various ellipsoidal forms. A cube can be distorted by pulling and extending the form along an axis, such as the diagonal between two opposite vertices, to provide various diamond-shaped forms, or transformed into a range of prismatic-type objects by extending or shortening its length, breadth or depth. Simple rectangular block prisms, cylinders, cones and pyramids can be distorted in numerous ways, including through compression, lengthening, pulling out and pushing in, as well as by altering the base or cross-section size or shape.

A process of cutting and re-assembly was considered by Wong in the *Principles of Three-Dimensional Design* (1977: 14–15). After introducing a cube, he showed how it could be dissected or deconstructed by cutting parallel cross-sections in lengthways (or breadthways or depthways) directions to produce a series of slices of equal size and shape, each with rectangular edges; alternatively, through diagonal cutting, slices of various shapes with bevelled edges could be produced (1977: 16). Examples are given also in Figure 8.16. Following from this dissection, Wong considered the position and direction of constituent slices of a lengthways-cut cube and showed how slices could be adjusted by rotation, curling, bending or further cutting and, after

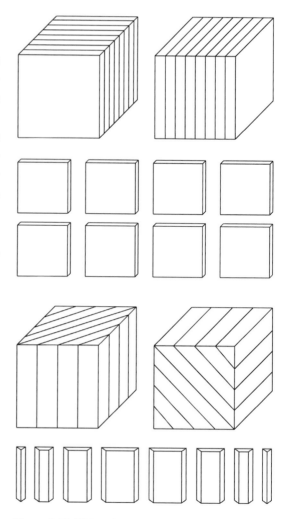

Figure 8.16 Slicing of a cube, lengthways, breadthways and diagonally (JC, adapted from Wong 1977: 16).

re-arranging, could offer several alternative three-dimensional forms. Wong showed how this process of slicing and adjusting the components of a cube and other three-dimensional forms could be fruitful in creating various small-scale models of interest to designers working in three dimensions. As indicated previously, digital versions of on-screen models were readily producable by the early-twenty-first century using commercially available computing technology.

Figure 8.17 Transformation of a prism through cutting and combining components (JC, adapted from Wolchonok 1959: 33).

Additive or subtractive transformations of commonly known forms offer further avenues for design development. As the terms imply, additive transformations involve adding one or more

components to a form, and subtractive transformations involve the removal of certain parts. By way of example, Figure 8.17 shows a prism cut into various related component parts (top row of figure) and a selection of forms created from assembling various of these cut components.

Subsequently to transformation, the relevant form may 'still retain its identity as a member of a family of forms' (Ching 1996: 48). Obviously, if the transformations are too substantial, the identity of the original form may be lost. So, provided that the cutting and re-arrangement, dimensional distortion, subtractive or additive transformations are small in size (say less than 50 per cent) compared to the size of the form being transformed, the visual integrity of the original form can be retained. With a cube, for example, relatively small adjustments can be made in its dimensions (length, breadth or depth), small components (such as smaller squares) can be cut from its faces or vertices, and additions of other forms (such as small hemispheres, pyramids or cones) can be made, and the original cubic form can still be recognizable.

An important point to stress is that no transformations should be the outcome of guesswork; rather, they should be based on the consideration of one of the proportioning systems mentioned in various chapters of this book. Such consideration will ensure visual order and consistent visual relationships between the various components of the design and between the components and the whole.

CHAPTER SUMMARY

The emphasis in this book has been largely on considering a range of fundamental design concepts and principles, most conveniently described, visualized and understood as resting in an imaginary two-dimensional world. However, it is important to realize that the vast bulk of designs, when manufactured, exist in the real three-dimensional world. Invariably, designers design on a two-dimensional plane (screen or sheet of paper). When the concern is with two-dimensional illustrative or surface designs (e.g. wallpapers, tilings and many textiles), a view from one direction is sufficient to understand the nature of what is intended. So, from a single representation on the sheet of paper or screen, it is not too difficult to envisage the intended finished product. When the intention is to design a three-dimensional object, visualization is much more complex, and several views may be required in order for an onlooker (or client) to comprehend the intentions of the designer. Early-twenty-first-century developments in computer technology have helped greatly in the communication of this more complex visualization.

An awareness of structure and form is of importance to all designers. In addition to various aesthetic concerns, designers working in three dimensions often need to consider practical aspects of performance of the intended final product. These considerations may, on the one hand, relate to the raw materials used or other aspects of the manufacturing process but, on the other hand, may relate closely to structure and form. Throughout much of the natural world, it is apparent that performance is governed closely by structure and form. This seems to be the case also in the designed and manufactured worlds. It may well be the case that underlying structural points which play an aesthetic role in a design may function also in a practical sense. For this reason, the concepts presented in previous chapters should prove of value to designers concerned with three-dimensional constructions.

Various three-dimensional geometric figures (including prisms, cylinders, cones and pyramids) were introduced in this chapter. It is important to note that such figures have occasionally acted as initiators of three-dimensional design forms. Also presented was a range of lattice structures (with dimensions based on golden-section and root-rectangle ratios). Attention was paid also to how various forms can undergo transformation, using processes referred to as distortion, cutting and re-assembly, addition and subtraction.

9

variations on a theme: modularity, closest packing and partitioning

introduction

The objectives of this chapter are to introduce a concept known as modularity and to review its applicability across a range of areas, including nature, art, design and architecture. Subsequently, the relationship between modularity and the concepts of closest packing and partitioning is explored, and a theory of emergence, whereby small component parts may interact with each other in such a way as to bring about unexpected results, where the sum of the whole is greater than the sum of the individual parts, is developed.

the nature of modularity

Modular design subdivides the design process into smaller manageable stages, each focused on the design of a module or component part of the intended object or construction. Individual modules are produced independently, are self-contained and can be slotted into the larger design solution. Lipson, Pollack and Suh (2002) presented an interesting paper concerned with modularity in biological evolution and remarked on the possible applicability to engineering design methods. Various research publications have reported on the use of modular systems in a range of industrial sectors: Cusumano (1991), Post (1997)

and Meyer and Seliger (1998) examined software development; Baldwin and Clark (1997) considered modularity in computer manufacture; Sanderson and Uzuneri (1997) focused on modularity in consumer electronic products; Cusumano and Nobeoka (1998) were concerned with the context of automobile manufacture; Worren, Moore and Cardona (2002) examined strategic aspects of modularity and product performance in the home appliance industries of both the United States and the United Kingdom.

In the treatise, *Structure in Nature is a Strategy for Design,* Pearce encapsulated the concept of modularity with the phrase 'minimum inventory—maximum diversity' (1990: xii), by which he meant systems of design with a few components which could be combined variously to yield a great range of forms. Modularity refers to the degree to which a set of parts within a design can be separated and recombined, with a mixing and matching of components. In other words, from a few basic elements (or modules), a large variety of possible structures (or solutions) is possible. Another feature of modularity is that components can be added and subtracted. The concept is applicable with only small degrees of variation across a wide range of contexts and can be detected throughout natural and human-made environments. In nature, modularity can refer to the expansion of a cellular organism through the addition of standardized units as is the case with the hexagonal cells of the honeycomb. Modularity offers great potential for innovation in the decorative arts and design, and is common in product design, surface-pattern design, architecture, interior design and furniture design, as well as in engineering. Several notable twentieth-century architects including Frank Lloyd Wright, Le Corbusier and Buckminster Fuller (who was, of course, an engineer as well as an architect) were enthusiastic proponents. Flexibility and ease of use are achieved by the modular approach. Instead of thinking in terms of a system as a collection of unconnected pieces, objects or parts (e.g. in the context of industrial design, switches, wires, rollers, cog wheels, circuits and cables), modularity aims for a collection of interrelated functional components or modules that can be arranged or rearranged with speed and efficiency, like children's building blocks. In fact, *Lego* bricks, used by generations of children across much of the world, are a powerful example of the concept.

Modularity can make potentially complex arrangements manageable, can be energy conserving, and is invariably cost-efficient. However, it is not without costs as the process of modularizing a complex design problem may be a lengthy process. Examples of modular systems in the modern era can be found in urban, domestic and working environments (e.g. buildings, kitchens and offices, respectively) and, from previous decades, among looms, railroad signalling set-ups, electrical power distribution systems and telephone exchanges. Modularity in design often combines the advantages of standardization (especially high volume and the resultant low manufacturing costs) with advantages associated with customization (giving each consumer the belief that she/he is receiving something unique); this is particularly the case following late-twentieth-century and early-twenty-first-century innovations in

computing and communications technology, which allowed manufacturers to respond exceedingly quickly to changes in consumer demand and in some cases permitted consumers to dictate which components (or modules) of a product they desired. Modularity can therefore enhance consumer choice in that bespoke (or made-to-order) items can be generated at short notice.

Modularity in automobile manufacture permitted certain components to be added or removed without necessitating further alterations to the basic design. By the early-twenty-first century, most automobile manufacturers worldwide offered a basic model in each range of products and the possibility of upgrades such as high-performance wheels, seasonal tyres, a more powerful engine, interiors in different materials with different seating arrangements and dashboards with various grades of hi-fi, entertainment and satellite navigation systems. All these add-ons were possible without change to the overall construction (including the chassis, exhaust and steering systems) of the basic model.

modularity in the fine and decorative arts

Modularity can allow alteration through reconfiguring, removing or adding constituent parts. In the fine arts this may take the form of joining together standardized units or modules to form larger and more complex compositions. Works of art permitting alteration through the rearrangement of components parts date at least from the European Renaissance. An example is the triptych *Garden of Earthly Delights* painted by Hieronymus Bosch (ca.1450–1516). Examples which embrace the concept to an even greater degree can be found in the twentieth century among the so-called modular constructivists. Modular constructivism was a style of sculpture which emerged during the 1950s and 1960s, based on the use of carefully selected modules which permitted intricate and in some cases numerous alternative combinations. The challenge for the artist was to determine the combinatorial possibilities of component parts. Screen-like forms, used in architectural contexts to divide space, filter light and add aesthetic interest, were an important development. Major proponents included artists such as Norman Carlberg and Erwin Hauer (both were students of Josef Albers, a key participant in the Bauhaus). Modularity was also considered to be an important aspect of the Minimalist art of the 1960s. Participating artists included Robert Rauschenberg, Dan Flavin, Donald Judd, Sol LeWitt and especially Tony Smith the sculptor, who began his career as an architectural designer and was deeply influenced by the work of Frank Lloyd Wright, a noted proponent of modularization in architecture. Modularity in the decorative arts has a long and diverse pedigree across historical and cultural contexts. A substantial review (1995) was provided by Slavik Jablan of the use of modularity in various Paleolithic ornaments from eastern Europe and central Asia, Roman mazes and labyrinths, various key patterns, ornamental brickwork, Islamic and Celtic knot designs and op-art examples from the late-twentieth century.

modularity in design and architecture

It is in the realms of design and architecture that the concept has held greatest influence; the nature of this influence and associated theoretical principles are explained briefly here. In the context of industrial design, modularity allows larger entities to be created from smaller component subentities, as well as the selection of interchangeable components in the manufacture of a product. Modularity can offer reduction in costs and flexibility in final design. It can combine the advantages of standardization (principally economies of scale) with those of apparent uniqueness from the viewpoint of the consumer (a stimulus for improved consumer demand). As noted previously, modularization in the modern era offers the possibility of manufacturing products from standardized units, in apparently unique configurations. This apparent uniqueness results from the personalized selection of component parts, and the belief that the ingredients selected combine uniquely to meet the exact specifications dictated by the consumer. The seemingly contradictory production modes of mass manufacture and personalized customization are thus combined (and may be referred to as mass customization), a development which has followed in the wake of innovations in computing and communications technology. By the early-twenty-first century, a vast range of product manufacturers as well as architectural, building and interior-design companies worldwide had recognized the commercial potential of mass customization; this could be seen clearly in the automobile and computing industries as well as in the building of apartment blocks, the lay-out of living accommodation and the manufacture of office, kitchen, bedroom and bathroom products across much of the economically developed world in the early-twenty-first century. Mass customization can also be associated with the early-twenty-first-century revolution in digital technology, which permitted seemingly unique combinations of access to entertainment, education or data processing features or the use of various software options, received through home- or work-based, hand-held or portable devices, all enabled through the purchase of a so-called unique package, consisting of units or modules or combinations selected by the consumer.

More often than not, brick-built constructions are created predominantly from multiples of one unit (i.e. a standard-size brick) and can therefore be considered as modular arrangements. Modularity has also entered the architectural realm in other ways. Architectural design can involve considering constituent parts of a building (e.g. its rooms) as modules, which can be added or subtracted at will. An office unit or residential apartment, for example, can be created to the specifications of the customer and can include the desired numbers, shapes and sizes of rooms suited to the client's needs and budget. Invariably, the modular components (walls, floors, ceilings, roofs etc.) are manufactured on an assembly-line basis, in a factory-type facility and delivered to, and assembled at, the intended site. Once assembled, modular-type buildings are largely indistinguishable from conventional, site-built constructions, but are invariably cheaper to build and cheaper to buy. Several advantages are apparent, including the elimination of construction delays due to inclement weather or late raw-material delivery; much less time spent on site prior to the completion of the building; ability to address the building needs of remote locations (i.e. those locations distant from established cities or areas of high population density), where costs of building on site are typically relatively high, raw materials are difficult to acquire and appropriately skilled labour is relatively scarce; lower quantities of waste on site; reduction of on-site disturbance caused by movement of vehicles, people and delivery of raw materials.

The prefabricated, factory-manufactured, mass housing units (known in Britain as prefabs), built in the decade following World War II, are an example of modularity. Typically, such constructions were built on an aluminium frame (though timber and steel varieties were made also) and had an entrance hall, two bedrooms, a living room, a bathroom (with a separate WC) and an equipped kitchen (with a basic stove, a few storage cupboards and an eating area), all contained in four or possibly five pre-built modules which could be transported to the intended site and assembled in a small proportion of the time taken using conventional, on-site construction. Another example of modularity, often cited by architectural historians, is the Hanna-Honeycomb House, also known as Hanna House, designed by Frank Lloyd Wright, using hexagon-shaped modules which gave the floor plan the appearance of a honeycomb. The modular construction allowed the building to be expanded and adapted over a twenty-year period as the needs and demands of the residents changed. Further potential features of modularity in architecture include efficient use of available space and economy in the use of materials. The concepts of closest packing and efficient partitioning are introduced in order to explore these features.

closest packing

The term *closest packing* is used to refer to circumstances where objects of identical shape and size are brought together in an efficient, stable

Figure 9.1 Honeycomb, image from Wikipedia, May 2011.

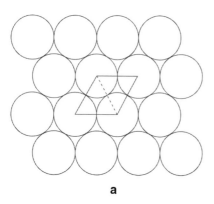

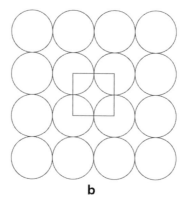

Figure 9.2a–b Closest packing with triangular formation and with less efficient square formation (JSS).

a   b

manner to ensure that the space between each is eliminated or kept to the minimum. Theories associated with closest packing have developed through considering the packing of equal-sized spheres or similar geometric figures. Many crystal structures are based on the closest packing of atoms. Closest packing finds expression in the three-dimensional ordering of polyhedron-shaped cells in biological systems as well as in the arrangement of atoms in certain metals (Pearce 1990: xiv). The bees' honeycomb (Figure 9.1) offers a familiar example of closest packing, with cells of regular-hexagon cross-section; the arrangement can hold the greatest amount of honey, using the least amount of wax and requiring the least amount of energy to be expended by the bees in the construction when compared to other polygon-shaped constructions.

The principle of closest packing is associated in particular with what is referred to as triangulation, as triangulated assemblies have an inherent stability, offer high strength per unit weight and are less energy-consuming than other arrangements in their construction and use. Closest packing using a triangular formation of equal circles in a given area is more efficient than a square formation (Figure 9.2a–b). Conservation of resources is part and parcel of a minimum-inventory/

maximum-diversity set-up (Pearce 1990: xiv). The principle of closest packing/triangulation is universal, operates at the nano-, micro- and macro-levels, and seems to establish conditions of minimum energy use. Good early examples are the three-dimensional triangulated structures (tetrahedral kites and space frames) designed by Alexander Graham Bell in early experiments associated with manned flight and the development of aircraft.

As observed by Pearce (1990: xiv), the form of objects is determined by two categories of fundamental forces: intrinsic and extrinsic forces. The first are those forces inherent in the make-up of the object (its physical component parts or chemistry); the second are those forces which come from the environment external to the object. Pearce (1990: xiv) cited the example of the snowflake: its molecular structure is a product of the intrinsic forces governing its form, while temperature, humidity and wind velocity are among the extrinsic forces influencing the same form. All forms in nature can be seen to result from the interaction of intrinsic and extrinsic forces.

efficient partitioning

An important aspect of closest packing and modularity is partitioning. The most efficient partitioning is the division of the available space into equal cells of maximum size for a given length of circumference. So the problem of partitioning could be stated as this: What is the most efficient way of dividing two-dimensional space with a network of equal-sized and equal-shaped cells so that the maximum area is contained in a cell with the smallest circumference? It is common knowledge that the circle, more than any other plane figure, encloses the greatest surface area for a given circumference or 'alternatively encloses a given area with the smallest circumference' (Pearce 1990: xiv). In the three-dimensional context, the sphere holds the greatest volume within the least surface area. Although both the circle and the sphere rank in the top positions when considering two- and three-dimensional space-filling (or closest packing) respectively, they are not as successful when it comes to being the most efficient means of partitioning space. When a group of circles are brought together, concave triangles are formed between surfaces; such triangles give the least area with the greatest circumference (the opposite of what is aimed for in efficient partitioning). However, if the circles are allowed to change their shape so that the concave triangles are removed, thus creating hexagons, a more efficient form of partitioning a surface into equal units of area can be created. These hexagonal cells match minimum structure (the overall wall length) to maximum usable area (Pearce 1990: 4).

Spheres do not offer the most efficient partitioning of three-dimensional space. When packed together, each sphere will have twelve spheres surrounding it (six around a central one, and with three on top and three on bottom). There is thus some space left unfilled. If, however, the sphere is allowed to expand its surface to fill that empty space, it can form a rhombic dodecahedron (Figure 9.3) which is more efficient than the sphere in partitioning three-dimensional space. These considerations of utilizing space are of particular importance to architects and interior designers. In *Structure in Nature is a Strategy for Design,* Pearce (1990) focused particular attention on the least-energy-using and high-efficiency structures and, as mentioned previously, presented an enquiry relating to the proposition that maximum diversity is obtainable from a minimum inventory. Although the perspective taken is that of the architectural practitioner and theoretician, all designers and visual artists could benefit greatly from considering this publication.

towards a theory of emergence in the decorative arts and design

The concept of emergence, adapted here from Holland (1998), can be used to consider how a collection of apparently simple components can

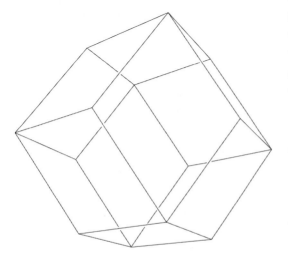

Figure 9.3 Rhombic dodecahedron (MV).

be brought together to form more complex enti-ties, a case where the sum of the whole is greater than the sum of the individual parts. Ant colonies, bee hives, snowflakes, termite hills, the internet and the global economy are good examples of emergence. The concept appears to operate occasionally in the decorative arts and design, and seems also to relate to concepts associated with modularity. In regularly repeating patterns, for example, a small component motif may, on its own, be relatively insignificant visually; but when associated with other copies of itself, the complete design that emerges can be read visually in various ways and can exhibit properties not apparent in the single original motif. This is largely because, in many regularly repeating patterns, visual connections between repeating units may only become apparent when the design is seen across the plane rather than as one small section. The viewer may make visual connections between various components within repeating units. Foreground or background effects and similarity or contrast between components within a design may be of importance.

CHAPTER SUMMARY

This chapter has been concerned with modularity, a concept of particular importance in architecture and design. The meaning of the concept is captured well in the phrase 'minimum inventory—maximum diversity' (Pearce 1990) suggesting that a vast range of possibilities can unfold from different combinations of a few ingredients. Examples of student responses to an assignment involving cutting regular polygons into several tiles, colouring these, making several copies and reassembling them into collections of repeating patterns (or tilings) are provided in Figures 9.4 and 9.5. A brief review is presented on the use of modularity in the decorative arts, design and architecture. Associated concepts such as closest packing and efficient partitioning have been explained briefly, and a theory of emergence in the decorative arts and design has been suggested.

Figure 9.4 Minimum-inventory and maximum-diversity tilings (Claira Ross).

Figure 9.5 Minimum-inventory and maximum-diversity tilings (Matthew Brassington).

structural analysis in the decorative arts, design and architecture

introduction

Knowledge of the geometric principles, concepts and perspectives underpinning structure and form in the decorative arts, design and architecture has its source in various ancient cultures, including dynastic Egypt, Assyria, ancient India and China as well as classical Greece and Rome. From the time of Euclid (ca. 300 BCE), geometry has been the tool of choice for architects, builders, artisans and designers. An understanding of various basic geometric principles (many identified and explained in this book) has offered generations of practitioners an avenue to address and solve problems relating to structure and form. Such understanding has also been of value in the visual analysis of designs and other products of creative endeavour. In view of this, the objectives of this chapter are first to review briefly a small selection of attempts by researchers to analyze structure and form in the decorative arts, design and architecture, and second to propose the basis for the development of a systematic analytical framework to be used by art-and-design analysts.

symmetry analysis—steps towards a consistent method

In the late-nineteenth and twentieth centuries, certain publications served as important milestones in the identification of the structural characteristics of regularly repeating two-dimensional designs. Meyer (1894 and 1957: 3) grouped such designs according to their spatial characteristics (e.g. enclosed spaces, ribbon-like bands or unlimited flat patterns, corresponding in the terminology employed in the present book to motifs, frieze patterns and all-over repeating patterns). Most importantly, he recognized that the two-dimensional forms exhibited in repeating all-over designs relied on an underlying structural network, thus anticipating the lattice networks used by scientific investigators and applied to the classification of crystallographic repeating patterns. As stated previously, in addition to exhibiting various symmetry characteristics, all regularly repeating all-over patterns are based on underlying grids or lattice structures (known as Bravais lattices, of which there are only five types). Stephenson and Suddards (1897: Chapters 2–5), in their examination of textile patterns, illustrated constructions based on rectangular, hexagonal, rhombic and square lattices. In a similar vein, Day (1903) stressed the importance of the underlying geometric structure of two-dimensional regularly repeating patterns and presented illustrations of patterns based on square, rhombic, parallelogram and hexagonal structural frameworks.

The twentieth century was witness to another related perspective on design analysis and classification based on examining the underlying symmetry characteristics of repeating designs. As noted previously, this means of analysis and classification was used in the latter part of the twentieth century to categorize two-dimensional regularly repeating patterns from various cultural contexts and historical periods. Important contributions were made by Washburn and Crowe (1988), Hargittai (1986 and 1989) and Hann (1992, 2003a). A key finding from the numerous research contributions was that, when a representative collection of patterns from a given cultural setting is analysed with respect to its underlying symmetry characteristics and classified into the various symmetry classes (seven border-pattern types and seventeen all-over-pattern types), it can be seen that different cultures express different symmetry preferences. An important implication when considering the distribution of symmetry classes in any representative series of data is that symmetry classification is a culturally sensitive tool, and can be used to detect continuity and changes over time (subject to the availability of suitable data). A review of relevant literature was provided by Hann and Thomson (1992). The important point to stress, from the viewpoint of hypothesis testing and theoretical development, is that symmetry analysis and classification allows for replication of results from one researcher to another (Washburn and Crowe 1988 and 2004).

geometric analysis in the decorative arts, design and architecture

Early attempts to analyse structure and form in the decorative arts, design and architecture can be traced to the Italian Renaissance (e.g. the work of Leone Battista Alberti); such analyses were focused invariably on understanding the inherent characteristics of Greco-Roman styles or orders of architecture and related forms of artistic endeavour. The stimulus was seemingly the belief that such sources held the geometric secrets of ancient Greek geometers, and that success was assured should such secrets be incorporated into the creative endeavours of fifteenth-century Italian practitioners (Chitham 2005: 20). Centuries later, Hogarth (1753) set out to identify the attributes of beauty and made some notable comments relating to structure and form; of particular importance is his view that humans have an inherent means of perceiving pleasing proportions in objects and that minor degrees of variation in proportions are more easily detected without measurement than changes in absolute dimensions of an object. Subsequent to Hogarth's observations, Owen Jones's *The Grammar of Ornament* (1856) stands as the great nineteenth-century treatise which assembled colour-printed reproductions of artistic output from a variety of cultural contexts and historical periods. The intention of Jones's work and many similar publications from the late-nineteenth and early-twentieth centuries (e.g. Racinet 1873; Speltz, 1915) was to identify the important stylistic attributes of decorative art and design from different sources, cultures and periods.

During the twentieth century there was much attention focused on the geometric analysis of (non-repeating) design also. Notable contributions in identifying important parameters which could possibly be of value to design analysts were made by Hambidge (1926), whose study of Greek art led him to propose that all harmonious design was based on what he termed *dynamic symmetry* and proportions provided by various root rectangles and golden-section-related whirling-square rectangles and related figures. Ghyka (1946) acknowledged the usefulness of Hambidge's perspectives, proposed a theory of proportion (which made reference to Greek and Gothic art) and discussed the relationships between geometry, nature and the human body. He also conducted geometric analyses on various examples of art and design from the late-nineteenth and early-twentieth-centuries. The most important published contributions to the developing knowledge relating to structure and form in art and design during the twentieth century did not in fact come from art or design practitioners or theorists, but rather from scholars concerned primarily with the geometric aspects of structures from the natural world. The works of Theodore Cook (1914) and D'Arcy Thompson (1917) are of note. In fact, the bulk of publications which set out to identify or examine aspects of structure in art and design make reference invariably to one or both of these texts.

By the early-twenty-first century, geometrical analysis of decorative art, design and architecture from different cultural or historical contexts had become a common research activity among mathematicians, and there was a plethora of relevant literature, the bulk of which was not readily accessible to typical art-and-design audiences. Among the exceptions were Melchizedek (2000), who examined cultural and historical aspects of what he called 'the flower of life' motif (consisting of overlapping circles in hexagonal order); Ascher (2000: 59), who discussed the challenge of including 'ethnomathematics' (defined as 'the study of the mathematical ideas of traditional peoples') in the school geometry curriculum; Elam (2001), who provided a brief review of the use of geometry in art and design from ancient times and presented summary analyses which included consideration of a selection of twentieth-century product designs and posters; Kappraff (2002), who in his treatise *Beyond Measure*

examined various geometric phenomena in cultural contexts, including aspects of the so-called Brunes Star and how it may have been used as a measure in ancient times; Gerdes (2003), who examined the emergence of mathematical thinking and expressed a central concern about developing a suitable methodology for the study of developments in early geometrical thinking; Fletcher (2004, 2005 and 2006), who presented a series of structural studies which included consideration of the *vesica piscis,* assemblies involving six circles around one and the golden section and its appearance within the regular pentagon and other geometric constructions; Marshall (2006), who examined manipulations of the square in an ancient Roman architectural context; Stewart (2009), who provided a review of the use of the Brunes Star (referring to it as 'a starcut diagram'). Importantly, a consistent, fully formed, replicable methodology and analytical framework to examine structural aspects of the decorative arts, design and architecture had not emerged. However, it should be noted that an important contribution in this direction had been made by Reynolds (2000, 2001, 2002 and 2003), who presented a range of useful articles dealing largely with procedures and other issues relating to the geometric analysis of designs. In his 2001 article he proposed a series of systematic stages to be followed by researchers when conducting geometric analysis in the decorative arts, design and architecture.

frequently used constructions and measures

Across the range of literature dealing with the structural aspects of the decorative arts, design and architecture, various geometric characteristics, principles, concepts, constructions, comparative measures, proportions and ratios are deemed of importance (in varying degrees) to art-and-design practitioners as well as analysts. These include the following: 1:1 (square); π:radius (circle); square-root series $\sqrt{2}$ (= 1.4142…):1; $\sqrt{3}$ (= 1.732…):1; $\sqrt{4}$ (= 2):1 and so on; regular polygons (particularly squares, pentagons and hexagons), Reuleaux polygons, the *ad quadratum,* the *vesica piscis,* various constructions associated with the sacred-cut square, the so-called Brunes Star; the golden section (Phi (Φ) or 1.618:1) and various associated constructions such as the golden rectangle or golden spiral; triangles (equilateral, isosceles, right-angled, scalene); lattice structures (including Bravais lattices) and grids based on the Platonic, Archimedean or other sorts of tilings; geometric symmetry and its component geometric operations (or symmetries). Many of these were listed by Reynolds (2001) in 'The Geometer's Angle: An Introduction to the Art and Science of Geometric Analysis', an article that has proved of great value to students interested in pursuing projects which set out to conduct structural analyses in art or design.

questions relating to method and data collection

Occasionally in the literature it is implied that ancient craftspeople possessed a range of mathematical tools largely beyond the comprehension of the bulk of accomplished modern-day practitioners. At this stage it is worth stressing that the author is not of the view that a vast range of geometric constructions were known to ancient artisans and builders. It is not denied, however, that sophisticated geometric knowledge and skill were possessed in ancient times (possibly among accomplished artists and architects in well-established centres of civilization). It seem more likely that craftspeople (those involved in building walls, placing bricks or tiles, cutting or carving wood, producing ceramics and weaving textiles) relied on artistic and creative skill and judgement, based on practical experience and innate ability, probably initiated or stimulated through past apprenticeships to more experienced practitioners. Also, it was probably the case that a limited number of tried and tested constructions were known to each group of ancient craftspeople. It is clearly the case historically (say, over the past millennium) that different regions developed diverse artistic or building styles, which may have resulted from local knowledge of a limited number of relevant geometrical constructions. What seems likely is that the different regions, and thus cultures, possessed their own unique combinations of geometrical knowledge, specific to the decorative arts, design and architecture of their locality. In order to determine whether or not this is indeed the case, it is necessary to examine representative series of data and to classify these data in a replicable and consistent manner. With this in mind, the intention here is to propose a systematic analytical framework to be used by researchers in the quest to conduct structural (i.e. geometrical) analyses in the decorative arts, design and architecture. The geometrical analysis of the type proposed by Reynolds (2001) is thus deemed to be worthy of development, while at the same time cognizance must be taken of the necessity for rigour in data collection and analysis stressed by scholars such as Huylebrouck (e.g. 2007).

steps towards a systematic analytical framework

Structural analysis in art and design can proceed in numerous ways. The stages proposed here are intended to form the basis of a more substantial analytical framework which, in the longer term, will contribute to the further understanding of how geometry has been used in the decorative arts, design and architecture historically and culturally. The procedures presented below assume that the analyst is unable to carry out direct measurements of the object itself and are intended for analysis of photographic or other images. If direct measurement of the object (say length, breadth and depth as well as the positions of key elements of the design) is indeed possible, then the derived data should be integrated with other data resulting from the photographic analysis. Common sense should prevail, and real measurements should take precedence over measurements conducted on photographic images should contradictions arise.

First, the item to be analyzed is selected and images of the object obtained (invariably in the form of drawings, paper photocopies or photographs). A clear and accurate image is selected. In the context of architectural representations, the principal façade seems to be the common-sense choice, though floor plans could be an equally enlightening source for analysis. With items of sculpture or product design, the view considered to be the front of the object is selected for measurement, though a side view could be examined as well. Importantly, before the analysis progresses, the analyst should identify key elements of the design: these could be the apex of a roof; the positioning of a rose window; the position of the entrance; the base or pinnacle of a spire; the positioning of columns or other supports; key points of focus in a piece of sculpture; key graphic characteristics of a poster or other two-dimensional representation or composition. In order to make the analysis valid it is important that these key elements are identified at this first stage.

Second, a rectangle should be drawn to frame, as closely as possible, the object represented in the image. Common sense should dictate the orientation of this rectangle with sides parallel to what the viewer perceives to be the mid-way line of the object.

A system of measurement is selected also; this may be millimetres, inches or any other units of measurement provided that consistency is retained subsequently. Once selected, a ratio of length to breadth of the represented object can be obtained through simple ruler measurement of the surrounding rectangle or alternatively by making direct measurement of the image of the object itself (using common sense to identify the widest dimension and the narrowest dimension). These measurements should then be recorded, and ratios calculated (simply by dividing a greater length by a lesser length) and then compared to the ratios associated with the following:

1. Square, with sides in the ratio of 1:1
2. Root-two rectangle, with sides in the ratio of 1:1.4142...
3. Root-three rectangle, with sides in the ratio of 1:1.732...
4. Root-four rectangle, with sides in the ratio of 1:2
5. Root-five rectangle, with sides in the ratio of 1:2.236...
6. Rectangle of the whirling squares (i.e. a golden-section rectangle), with sides in the ratio of 1:1.618...

The selection of a particular rectangle (from the six categories given above) can be based on a percentage deviation or allowance of 2 per cent above or below the value given. Thus, if a ratio of say 1:1.43 was obtained, this result would fall within acceptable boundaries, and the rectangle could be classified as a root-two rectangle.

Third, with reference to the circumscribing rectangle, two corner-to-corner diagonals are drawn and the centre of the rectangle thus identified. Each angle at the centre is bisected and lines are drawn to connect the mid-points of opposite sides, thus dividing the rectangle into four equal parts (Figure 10.1).

Fourth, the eight half-diagonals are drawn from each angle to the mid-point of the two opposite sides (Figure 10.2). The attentive reader will realize that a Brunes-Star-type outline has been created (though probably on a non-square rectangle). This same underlying structure is commonly found on designs worldwide (e.g. various Islamic tilings follow the format). A wide-ranging review identifying the use of the structure in numerous contexts was provided by Stewart (2009).

As suggested previously, the Brunes-Star-type diagrammatic form identifies various points (where lines within the drawn rectangle overlap or meet at a point); these may be referred to as key aesthetic points (or KAPs), that is points where key elements of a design, composition or other construction may have been placed. In the context of this chapter, the Brunes Star has simply been adopted as a means of analytical measurement. For the sake of analysis, each point (twenty-five in total) within the Brunes-Star-type diagram can be given an arbitrary letter (A, B, C, D etc. up to Y, as shown in Figure 10.3).

The fifth stage of analysis is to determine whether any of the key elements of the design (identified at the first stage of analysis outlined earlier) are located at KAPs in the Brunes-Star-type diagram (drawn over the selected image in the previous stages). If so, the location of each within the rectangle should be recorded by reference to

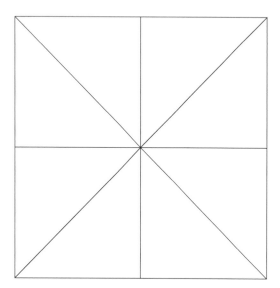

Figure 10.1 Brunes-Star construction, first stage (AH).

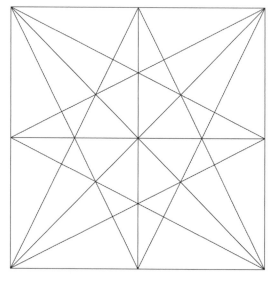

Figure 10.2 Brunes-Star construction (AH).

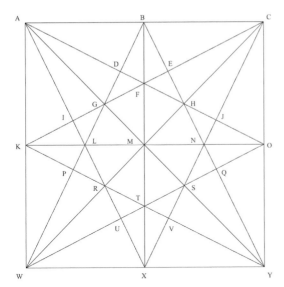

Figure 10.3 Brunes-Star construction, showing key aesthetic points (AH).

the letter designation (A, B, C, D etc. up to Y) suggested previously.

As stated in the section 'symmetry analysis—steps towards a consistent analysis', it is now established that symmetry analysis of repeating patterns may be used to highlight cultural or historical continuity, adaptation or change (see for example Washburn and Crowe 1988; Hann and Thomson 1992; Hann 1992). With this in mind, it seems sensible to propose that a geometrical analysis of the type described in these stages should be used as the basis on which to examine representative samples of designs from different sources. It is accepted that this proposal is at a rudimentary stage, and substantial development, adaptation and adjustment are required in order to produce a fully formed analytical tool. It is felt, however, that if the geometrical structure of designed objects, compositions or constructions is culturally sensitive, then the framework proposed should pick up differences from culture to culture

and from historical period to historical period. So, the next stage of development will be to test whether the series of measures proposed is culturally sensitive in the same way that symmetry analysis is known to be.

questions of accuracy

Often, for simple logistical reasons, accurate direct measures from the selected object, construction or building are not possible and reference must be made to photographic representations of one kind or another. As stated in the previous section, where accurate direct measurement (e.g. of length, breadth and depth) can be made and the positions of key elements of the design thus determined, these should indeed be made and integrated with other data resultant from the photographic analysis. The procedures proposed above, however, are intended for the analysis of photographic evidence. It is best if such photographs be taken by the investigator, but again this may not be possible. Analysis of photographic representations is prone to problems, some of which were highlighted by Reynolds (2001). Care must be taken in the selection of photographs. Perfectly frontal images are required. Images obtained from the internet may be distorted and may thus show greater or lesser length to breadth than the reality. In conditions where the exact provenance of photographs is unknown and dependability not assured, it would be best to collect as many versions as possible and to establish whether there is an acceptable average with an acceptable percentage of deviation from the two extremes. It is important that investigators do not step outside the specified margin of error (the allowance when classifying the rectangle type); investigator inaccuracy often hides behind so-called workman error or some past natural disaster. Differences can of course be expected between the conception of the design as a drawing and its realization in physical form, but the specified leeway should be sufficient to cater for such inaccuracy. If the intention is to make any form of comparison between series of data from different cultural or historical sources, it is crucial that the data are deemed representative for the source in question.

CHAPTER SUMMARY

A few key concepts and principles outlined in this book are suggested for incorporation in an analytical framework suited for analyzing and classifying two- and three-dimensional designs and allowing replication of results from one analyst to another. Designs may be created and developed by reference to structural rules, and, subsequently, they may be analyzed with respect to their underlying structural characteristics. The geometric analysis of representative groups of designs holds the potential to uncover a wide range of social, psychological, philosophical and cultural properties or characteristics, a potential suggested by the remarkable strides achieved in the related area of symmetry classification pioneered largely by the endeavours of scholars such as Washburn and Crowe (1988).

11

a designer's framework

Many theoretical principles and concepts have been covered in this book, and the intention of this short concluding chapter is not to restate these but rather to present a reiteration of proposals which may prove of some direct applicability to designers and other creative practitioners in addressing relevant design issues and solving design problems relating to structure and form. The focus in this concluding chapter is therefore on practical application and synthesis rather than on theory building and analysis.

Although the emphasis across the majority of chapters in this book has been on providing a framework to enhance design in the two-dimensional plane, the proposals, concepts or principles covered are still of potentially great value to practitioners concerned with three-dimensional design. It is accepted, however, that substantial further research and scholarship are required in order to make some of the content more precisely applicable across the full spectrum of design activity. It is hoped that future authors will take up this challenge.

Judgement of whether or not two or more components of a design are harmonious and work together visually is obviously based on subjectivity (though this subjectivity is often tempered by visual judgment gained from experience). The term *geometric complementarity* can be applied to circumstances where geometrical figures (both two-dimensional and three-dimensional) are in proportion to one another and are judged to express harmony, although they may be of unequal size, area, volume or dimensions. Geometric proportion is an objective measure, and its consideration may assist in deliberations relating to harmony between individual elements of a design.

The potential value of the series of root rectangles and the whirling-squares rectangle as compositional aids has been noted, and it has been proposed that designers or other creative practitioners consider the selection of one of these formats when creating a composition. In addition, it has been proposed that each rectangle can be usefully employed as a positioning device by the practitioner in decisions relating to the placement of important elements within a design. Steps of construction of the Brunes Star have been considered and applied to each rectangle in the series. Intersecting lines within each figure, as well as lines meeting at the same point on the sides, have been identified as key aesthetic points (or KAPs) and have been proposed as ideal locators for key elements of a design or other visual composition.

The importance of grids to designers and other practitioners in the visual arts has been highlighted several times (especially in Chapters 2, 3 and 6). It has been seen also that various classified tilings can be employed usefully as grids. The use of the equilateral-triangle and square grids (both Platonic tilings) to guide the construction of regularly repeating patterns is long-established historically. It has been proposed that consideration be given to the use of other classified tilings as grids, especially the semi-regular and demi-regular tilings (Chapter 4). In Chapter 4 a series of procedures has been proposed by which original tiling designs, based largely on manipulating the constituent cells of common classified tilings, can be created. A series of grids, based on various root rectangles and the whirling-squares rectangle, has been presented (Chapters 3 and 6). Various further manipulations, developments and additions are of course possible, and it is hoped that the reader will indeed make these and add them to the framework proposed here. Of great importance is the realization that design is a

systematic process, and that addressing problems relating to structure and form in design, and in the other visual arts in general, should be approached systematically. The structures which underpin all successful designs are invariably hidden from view in the finished product, but these are none the less crucial to the success of the product itself. There are, of course, exceptions in some modern buildings where, for example, aspects of the structural framework are incorporated within the final construction and may form an aesthetic feature (e.g. Figure 11.1). The principles and concepts presented in this book cover most of the ground required to underpin fruitful studio practice and resulting successful design.

Figure 11.1 Structure is best considered as the underlying, often hidden, framework supporting form (photo taken in central Seoul, Korea, 2007).

appendix 1

sample assignments and exercises

definitions and illustrations

You are required to present ONE definition and ONE original black-and-white illustration for EACH of the following visual elements/concepts:

(1) Point. (2) Line. (3) Plane. (4) Structure. (5) Form.

Your response should be suited for publication in a textbook for use by undergraduate students registered on art or design courses. Each definition should be in clear, simple and understandable English, and each illustration should be presented to professional, publishable standards. Original black-and-white photographic images (taken by you) may be incorporated in your submission if desired. You are required to submit all definitions and illustrations in both hard copy and digital form.

Limitations: each definition should be of thirty words or (preferably) fewer, and the total response should be presented within two sides of standard-sized paper (e.g. A4 size). Other than the brief definitions, further explanatory text should not be included. A simple, one-word heading may be included with each illustration if you believe that this is necessary for identification purposes. Text should therefore be kept to a minimum. It is crucial that all definitions and/or illustrations submitted be your own work. Simply downloading a definition or illustration from a Web site or other published source is not acceptable.

structural analysis in art and design

You have been commissioned to produce a two-page article for publication in the December edition of the (fictitious) quarterly journal *Visual Analysis and Synthesis*. The journal is published in English, and the principal target readership is European and North American undergraduate and postgraduate design students. Sales in Asia, Australasia, South America and Africa are increasing.

You are required to select a famous/renowned work of art or design from a period of your choice. This may be a painting, a building, an advertisement, a consumer product or any other designed product. Your selected object/image should be reproduced as a half-page illustration (titled and sourced) and presented as a component within your two-page article. Additional secondary illustrations may also be included. Font size must be no smaller than ten point. A generous margin is expected.

Within the specified two-page limit, you are required to provide a well-focused and informative geometrical analysis of your selected image/object. You are advised, where appropriate, to make close reference to the structural elements and principles dealt with in the present book. An appropriate title should be given to your article. Numbered section headings should provide a readable, accessible structure. All factual information and views taken from other analysts, theorists or commentators must be properly acknowledged and referenced. Assessment will be based on evidence of scholarship, clarity and quality of argument, evidence of further reading and understanding of relevant concepts and principles. Readability, quality of presentation and lay-out are further considerations. The work should be convincing as a publishable article. Close adherence to the specifications stated above, including full referencing of sources, is crucial.

modular tiling (cut, colour, rearrange and repeat)

This assignment is concerned with the production of two collections of tiling designs, produced using a small number of tiling shapes cut from a regular polygon.

The requirements: you are required to produce a professionally finished and presented collection of twelve patterns (or tiling designs), each created from tiling elements cut or drawn from a regular polygon (six designs from elements of a square, and six designs from elements of either an equilateral triangle or a hexagon). The process of creating the individual tiling elements is outlined in 'creating the patterns'. A minimum of four repeating units of each design must be shown. You must confine your total presentation of all twelve designs plus other necessary components of the assignment, detailed in the next paragraph, to two sides of a standard sheet of paper (e.g. A4 size). You must state and illustrate an anticipated end use and scale for your designs. End uses could include either exterior tilings or paving for walls, pathways or gardens (domestic, corporate or municipal) or interior uses such as floor or wall tilings, printed textiles or other surface designs for furnishings, carpets or other interior end uses. All of these are, however, obvious choices, so try to suggest something more profound and thus show your capability for original thinking. A process or means of application (or production) as well as the raw materials to be used must be stated also. Costs of raw materials as well as the costs of realizing one of your designs in your specified end use should be given.

All these criteria are normal requirements in professional design practice. The intention of this assignment is to develop your awareness of structure and form. For that reason, you are required also to include samples of design sources (mood or theme illustrations), the colour palette/s used to colour your designs and an end-use illustration. The design collection with all associated information and illustrations should be professionally presented. So, as a first step, decide upon an end use, the materials to be used, the scale of the intended design and the sources of inspiration (from which a colour palette of up to six colours can be selected, and ideas relating to texture as well as the shapes of the tiling elements can be developed).

Creating the patterns: draw a square to dimensions of your choice. Making reference to the forms expressed in your sources of inspiration, cut the square into three or more parts. Colour each tile with a colour of your choice. Do not feel restricted to the use of flat colour; textural qualities may be applied also. Make multiple copies (by scanning or photocopying each coloured tile). Use these three or more differently shaped tiles (in any numerical proportion you wish) to create a collection of six periodic tilings, which cover the plane without gap or overlap (this bit is difficult, so careful planning is required prior to cutting the polygon). Each design must be original, precisely drawn, distinctly different and must not rely solely on a change of scale as a means of differentiation. Feel free to use computing software of your choice. In order to ensure maximum variation, you must manipulate the cut tiles rather than the original square.

Repeat the process using either a regular hexagon or an equilateral triangle. Draw up and cut your selected polygon into three or more tiles. Colour each tile (using the same palette used previously). Make multiple copies and produce a second collection of six original, precisely drawn, distinctly different patterns. Ensure that each of your twelve designs is presented in a rectangular or square format, and shows a minimum of four repeats. You must also indicate how you cut up the two original polygons (i.e. what shapes of tiles were used for each of the two sets of six designs). The total presentation must be within the confines of two sides of a standard-sized sheet of paper (e.g. A4).

a publishing opportunity

A monograph series concerned with structure and form in design will be published in the future and will be aimed specifically at undergraduate art and design students. Relevant titles are as follows:

1. Islamic Tilings
2. Polyhedra and Other Three-dimensional Forms
3. Structural Analysis of Images
4. Modularity
5. Chaos and Fractals
6. Grids, Lattices and Networks
7. Symmetry and Patterns
8. The Golden Section, Rectangle and Spiral
9. Point and Line to Plane

Each monograph is produced to quarto size, with overall size 189 x 244 mm and a live text area of about 150 mm x 200 mm (i.e. allowing for margins).

Select four titles from this list. For each selected title, you are required to produce one original black-and-white, single-page image appropriate for use as a cover. Each image should be printed to the following maximum dimensions: 150 mm x 200 mm. No lettering of any kind should be included within the image itself (it should thus be clear to the assessor via the design/art work which image refers to which monograph title). Original black-and-white photographs may be included as responses to no more than two of your four selections.

beauty in the eye of the beholder

Select a frontal photograph of a human face that you consider to be very attractive. You are required to conduct a structural (i.e. geometrical) analysis of the image in order to ascertain the presence of notable proportions/ratios or other geometrical characteristics (e.g. ratios associated with Fibonacci-series numbers or various rectangles and other figures as well as reflection symmetry). It may be helpful to produce an enlarged, photocopied version of the image in order to facilitate ease of measurement. You may wish to conduct the following measures: top of head to tip of chin; centre of mouth to tip of chin; centre of mouth to tip of nose; tip of nose to bridge of nose; bridge of nose to pupil of eye; pupil to pupil; width of nose at outer part of nostrils; pupil to eyelash; eyelash to eyebrow; eyebrow to eyebrow; any other measure you believe to be appropriate.

You may then wish to establish if there are any apparent relationships between these measures. You may also wish to draw what you consider to be a mid-way line from head to chin to assess the image for reflection symmetry.

You are required to present your data in tabular form, indicating clearly which measures refer to which features. You are required also to make a concise, unambiguous and well-focused concluding statement/comment of no more than 200 words which summarizes in text form the results and conclusions to your mini-survey/experiment.

fractal images

Consider the fractal images depicted in Chapter 5. They exhibit self-similarity or scale symmetry. You are required to produce a series of original images which show the stages in developing a fractal image of your own, showing the first stage (the initiator, from which everything else develops) and four stages of iteration.

designer grids

Select a regular grid consisting of one of the following: equilateral triangles, squares or regular hexagons. Using software of your choice, carry out manipulations which result in the design of ten non-regular grids (with individual cells in each case of a different size, shape and orientation to each other as well as to the original). Your focus should be on producing a series of grids which you believe would be of value to a designer.

colouring polyhedra

Suppose you are given the task of colouring each of the Platonic solids. A requirement is that two faces which share an edge are not permitted to be the same colour. What is the minimum number of colours required for each of the five Platonic solids?

symmetry of motifs

Take ten photographs of motifs or other figures from your everyday environment. Making reference to the symmetry notations and illustrations given in Chapter 5, classify each of your ten images with respect to their symmetry characteristics.

polygons and their constructions

In the following exercises you are permitted to use any geometric instruments and drawing implements of your choice.

1. Draw an equilateral triangle.
2. Draw a square.
3. Draw a regular pentagon.
4. Draw a regular hexagon.
5. Draw a regular heptagon.
6. Draw a regular octagon.
7. Draw a regular enneagon.
8. Draw a regular decagon.

periodic tiling of the plane

1. Explain why there are only three Platonic (or regular) tilings and provide a clear, precisely drawn illustration of each together with an appropriate notation.
2. Present a precisely drawn illustration for each of the eight Archimedean (or semi-regular) tilings.

symmetry in border patterns

Photograph, copy or draw twenty border patterns and twenty all-over patterns from any published or observed sources. Making reference to the schematic illustrations and notations provided in Chapter 5, identify the constituent symmetry characteristics of each and classify each with the appropriate notation.

divide and reassemble

Figure A1.1 shows a rectangular prism with a section removed. The removed section is divided further, first into two components and then into three components. Make a selection of components (as many or as few as you wish from the four images presented in the illustration) and assemble these parts to produce five different constructions each of the same volume.

Repeat the process for each of the five Platonic solids (depicted previously in Figure 7.1), thus removing a section from each and cutting this removed section into two component parts and then into three component parts. In each of the five cases, produce five different constructions (each of the same volume) from the cut components.

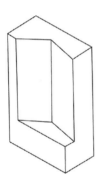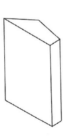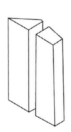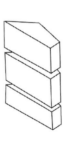

Figure A1.1 Divide and reassemble (MV, after Wolchonok 1959: 32).

appendix 2

explanation of pattern notation

explanation of frieze pattern notation

Frieze patterns exhibit regular translation in one direction only, and there are seven distinct classes. Various types of notation can be found in the relevant literature. The most commonly accepted notation is of the form pxyz; this gives a concise and easily understandable indication of the symmetry operations used in each of the seven classes.

The first letter p of the four-symbol notation prefaces all seven frieze patterns and is consistent with similar notation used for classifying all-over patterns. Symbols in the second, third and fourth positions (i.e. represented by x, y and z) indicate the presence (or absence) of vertical reflection, horizontal reflection or glide reflection and two-fold rotation respectively. The letter x, at the second position in the notation, will equal m if vertical reflection (perpendicular to the longitudinal axis) is present; otherwise, x will equal the number 1. The letter y, at the third position in the notation, will equal m if longitudinal reflection (parallel to the sides of the border) is present, will equal the letter a if glide reflection is present and will equal the number 1 if neither is present. The letter z, at the fourth position in the notation, will equal the number 2 if two-fold rotation is present and the number 1 if no rotation is present. The seven classes of frieze patterns can thus be classified as follows: p111, p1a1, pm11, p1m1, p112, pma2 and pmm2.

explanation of all-over pattern notation

There are various notations which have been used by mathematicians and crystallographers in the classification of all-over patterns. Although different from border patterns, a similar (though slightly more complicated) four-symbol notation, of the form pxyz or cxyz, is used. This indicates the type of unit cell, the highest order of rotation and the symmetry axes present in two directions. The role of each symbol is summarized in the following paragraphs.

The first symbol, either the letter p or c, indicates whether the lattice cell is primitive or centred. Primitive cells are present in fifteen of the all-over pattern classes and generate the full pattern by translation alone. The remaining two classes of all-over patterns have cells of the rhombic lattice type, with the enlarged cell containing two repeating units, one contained within the centred cell, and another in quarters of the enlarged cell corners.

The second symbol, x, denotes the highest order of rotation present. Where rotational symmetry is present, only two-, three-, four- and six-fold orders of rotation are possible in the production of two-dimensional plane patterns, as figures cannot repeat themselves around an axis of five-fold symmetry. This is known as the crystallographic restriction and was discussed by Stevens (1984: 376–90). If no rotational symmetry is present, x = 1.

The third symbol, y, represents a symmetry axis normal to the x-axis (i.e. perpendicular to the left side of the unit cell): m (for mirror) indicates a reflection axis, g (for glide) indicates a glide-reflection axis and 1 indicates that no reflection or glide-reflection axes are present normal to the x-axis.

The fourth symbol, z, indicates a symmetry angle at angle α to the x-axis, with α dependant on x, the highest order of rotation (shown by the second symbol). Angle α = 180 degrees if x = 1 or 2, α = 45 degrees if x = 4 and α = 60 degrees if x = 3 or 6. The symbols m and g denote the presence of reflection and glide-reflection symmetry, as with the third symbol. The absence of symbols (or the use of the number 1) in the third or fourth position indicates that the pattern has neither reflection nor glide reflection.

references

Abas, S. J. (2001), 'Islamic Geometrical Patterns for the Teaching of Mathematics of Symmetry', *Symmetry in Ethnomathematics,* 12/1–2: 53–65.

Abas, S. J., and Salman, A. S. (1995), *Symmetries of Islamic Geometrical Patterns,* Singapore, Hackensack, NJ, London and Hong Kong: World Scientific.

Allen, J. (2007), *Drawing Geometry,* Edinburgh: Floris Books.

Arnheim, R. (1954 and 1974), *Art and Visual Perception,* Berkeley: University of California Press.

Ascher, M. (2000), 'Ethnomathematics for the Geometry Curriculum', in C. A. Gorini (ed.), *Geometry at Work: Papers in Applied Geometry, MAA Notes,* 53: 59–63.

Baldwin, C., and Clark, K. (1997), 'Managing in an Age of Modularity', *Harvard Business Review,* 75/5: 84–93.

Baumann, K. (2007), *Bauhaus Dessau: Architecture, Design Concept = Architektur, Gestaltung, Idee,* Berlin: Jovis.

Bourgoin, J. (1879), *Les éléments de l'art arabe: le trait des entrelacs,* Paris: Fermin-Didot.

Bourgoin, J. (1973), *Arabic Geometric Pattern and Design,* New York: Dover.

Bovill, C. (1996), *Fractal Geometry in Architecture and Design,* Boston: Birkhauser.

Broug, E. (2008), *Islamic Geometric Patterns,* London: Thames and Hudson.

Brunes, T. (1967), *The Secret of Ancient Geometry and Its Uses,* 2 vols, Copenhagen: Rhodos.

Castéra, J. M. (1999), *Arabesques: Decorative Art in Morocco,* Paris: ACR edition.

Calter, P. (2000), 'Sun Disk, Moon Disk', in C. A. Gorini (ed.) *Geometry at Work: Papers in Applied Geometry, MAA Notes,* 53:12–19.

Chavey, D. (1989), 'Tilings by Regular Polygons II. A Catalog of Tilings', *Computers, Mathematics and Applications,* 17/1–3: 47–165.

Ching, F.D.K. (1996), *Architecture, Form, Space and Order,* New York: John Wiley & Sons.

Ching, F.D.K. (1998), *Design Drawing,* New York: John Wiley & Sons.

Chitham, R. (2005), *The Classical Orders of Architecture,* 2nd edn., Amsterdam: Elsevier.

Christie, A. H. (1910), *Traditional Methods of Pattern Designing,* Oxford: Clarendon Press, republished (1969) as *Pattern Design. An Introduction to the Study of Formal Ornament,* New York: Dover.

Cook, T. A. (1914), *The Curves of Life,* London: Constable, republished (1979) New York: Dover.

Corbachi, W. K. (1989), 'In the Tower of Babel: Beyond Symmetry in Islamic Designs', *Mathematics Applications,* 7: 751–89.

Coxeter, H.S.M. (1961), *Introduction to Geometry,* New York: John Wiley & Sons.

Critchlow, K. (1969), *Order in Space: A Design Sourcebook,* London: Thames and Hudson.

Critchlow, K. (1976), *Islamic Patterns,* London: Thames and Hudson.

Cusumano, M. (1991), *Japan's Software Factories: A Challenge to U.S. Management,* New York: Oxford University Press.

Cusumano, M., and Nobeoka, K. (1998), *Thinking beyond Lean,* New York: Free Press.

Davis, P. J. (1993), *Spirals: From Theodorus to Chaos,* Wellesley, MA: A. K. Peters.

Day, L. F. (1903), *Pattern Design,* London: B. T. Batsford, republished (1999) New York: Dover.

Doczi, G. (1981), *The Power of Limits: Proportional Harmonies in Nature, Art and Architecture,* Boulder, CO: Shambhala.

Dondis, D. A. (1973), *A Primer of Visual Literacy,* Boston: Massachusetts Institute of Technology.

Dürer, A. (1525), *Underweysung der Messung,* Nürnberg: Hieronymus Andreas Formschneider.

Edwards, E. (1932), *Dynamarhythmic Design,* New York: Century, republished (1967) as *Pattern and Design with Dynamic Symmetry,* New York: Dover.

Elam, K. (2001), *Geometry of Design: Studies in Proportion and Composition,* New York: Princeton Architectural Press.

El-Said, I., and Parman, A. (1976), *Geometric Concepts in Islamic Art,* London: World of Islam Festival Publications.

Falbo, C. (2005), 'The Golden Ratio: A Contrary Viewpoint', *College Mathematics Journal,* 36/2: 123–34.

Field, R. (2004), *Geometric Patterns from Islamic Art and Architecture,* Norfolk (UK): Tarquin.

Fischler, R. (1979), 'The Early Relationship of Le Corbusier to the Golden Number', *Environment and Planning,* 6: 95–103.

Fischler, R. (1981a), 'On the Application of the Golden Ratio in the Visual Arts', *Leonardo,* 14/1: 31–2.

Fischler, R. (1981b), 'How to Find the "Golden Number" without Really Trying', *Fibonacci Quarterly,* 19: 406–10.

Fletcher, R. (2004), 'Musings on the Vesica Piscis', *Nexus Network Journal,* 6/2: 95–110.

Fletcher, R. (2005), 'Six + One', *Nexus Network Journal,* 7/1: 141–60.

Fletcher, R. (2006), 'The Golden Section', *Nexus Network Journal,* 8/1: 67–89.

Gazalé, M. J. (1999), *Gnomon,* Princeton: Princeton University Press.

Gerdes, P. (2003), *Awakening of Geometrical Thought in Early Culture,* Minneapolis: MEP.

Ghyka, M. (1946), *The Geometry of Art and Life,* New York: Sheed and Ward, republished (1977), New York: Dover.

Gombrich, E. H. (1979), *The Sense of Order. A Study in the Psychology of Decorative Art,* London: Phaidon.

Goonatilake, S. (1998), *Towards a Global Science,* Bloomington: Indiana University Press.

Grünbaum, B., Grünbaum, Z. and Shephard, G. C. (1986), 'Symmetry in Moorish and Other Ornaments', *Computers and Mathematics with Applications,* 12B: 641–53.

Grünbaum, B., and Shephard, G. C (1987), *Tilings and Patterns,* New York: W. H. Freeman.

Grünbaum, B., and Shephard, G. C. (1989). *Tilings and Patterns: An Introduction,* New York: W. H. Freeman.

Haeckel, E. (1904), *Kunstformen der Natur,* Leipzig and Vienna: Bibliographisches Institut, reprinted (1998) as *Art Forms in Nature,* New York: Prestel-Verlag.

Hahn, W. (1998), *Symmetry as a Developmental Principle in Nature and Art,* Singapore: World Scientific.

Hambidge, J. (1926, 1928 and 1967), *The Elements of Dynamic Symmetry,* New York: Dover.

Hammond, C. (1997), *The Basics of Crystallography and Diffraction,* Oxford: International Union of Crystallography and Oxford University Press.

Hankin, E. H. (1925), *The Drawing of Geometric Patterns in Saracenic Art,* Mémoires of the Archaeological Survey of India, Calcutta: Government of India, Central Publications Branch.

Hann, M. A. (1991), 'The Geometry of Regular Repeating Patterns', PhD thesis, University of Leeds.

Hann, M. A. (1992), 'Symmetry in Regular Repeating Patterns: Case Studies from Various Cultural Settings', *Journal of the Textile Institute,* 83/4: 579–90.

Hann, M. A. (2003a), 'The Fundamentals of Pattern Structure. Part III: The Use of Symmetry

Classification as an Analytical Tool', *Journal of the Textile Institute,* 94 (Pt. 2/1–2): 81–8.

Hann, M. A. (2003b), 'The Fundamentals of Pattern Structure. Part II: The Counter-change Challenge', *Journal of the Textile Institute,* 94 (Pt. 2/1–2): 66–80.

Hann, M. A. (2003c), 'The Fundamentals of Pattern Structure. Part I: Woods Revisited, *Journal of the Textile Institute,* 94 (Pt. 2/1–2): 53–65.

Hann, M. A., and Thomas, B. G. (2007), 'Beyond Black and White: A Note Concerning Three-colour-counterchange Patterns', *Journal of the Textile Institute,* 98/6: 539–47.

Hann, M. A., and Thomson, G. M. (1992), *The Geometry of Regular Repeating Patterns,* Textile Progress Series, vol. 22/1, Manchester: Textile Institute.

Hargittai, I. (ed.) (1986), *Symmetry: Unifying Human Understanding,* New York: Pergamon.

Hargittai, I. (ed.) (1989), *Symmetry 2: Unifying Human Understanding,* New York: Pergamon.

Heath, T. (1921), *History of Greek Mathematics. Vol. I: From Thales to Euclid,* and *Vol. II: From Aristarchus to Diophantus,* Oxford: Clarendon Press, republished (1981) New York: Dover.

Heath, T. (1956 edn), *Euclid: The Thirteen Books of the Elements,* 3 vols., New York: Dover.

Hemenway, P. (2005), *Divine Proportion: Φ (Phi) in Art, Nature and Science,* New York: Sterling.

Holden, A. (1991), *Shapes, Space and Symmetry,* New York: Dover.

Holland, J. H. (1998), *Emergence: From Chaos to Order,* Oxford: Oxford University Press.

Huntley, H. E. (1970), *The Divine Proportion: A Study in Mathematical Beauty,* New York: Dover.

Huylebrouck, D. (2007), 'Curve Fitting in Architecture', *Nexus Network Journal,* 9/1: 59–65.

Huylebrouck, D. (2009), 'Golden Section Atria', Four-page manuscript provided courtesy of the author.

Huylebrouck, D., and Labarque, P. (2002), 'More Than True Applications of the Golden Number', *Nexus Network Journal,* 2/1: 45–58.

Itten, J. (1963 and rev. edn. 1975), *Design and Form: The Basic Course at the Bauhaus,* London: Thames and Hudson.

Jablan, S. (1995), *Theory of Symmetry and Ornament,* Belgrade: Mathematical Institute.

Jones, O. (1856), *The Grammar of Ornament,* London: Day and Son, reprinted (1986) London: Omega.

Kandinsky, W. (1914), *The Art of Spiritual Harmony,* London: Constable, republished as *Concerning the Spiritual in Art,* trans. M.T.H. Sadler (1977), New York: Dover.

Kandinsky, W. (1926), *Punkt und Linie zu Fläche,* Weimar: Bauhaus Books.

Kandinsky, W. (1979), *Point and Line to Plane,* New York: Dover.

Kaplan, C. S. (2000), 'Computer Generated Islamic Star Patterns', in R. Sarhangi (ed.), *Proceedings of the Third Annual Conference, Bridges: Mathematical Connections in Art, Music and Science,* Kansas: Bridges.

Kaplan, C. S., and Salesin, D. H. (2004), 'Islamic Star Patterns in Absolute Geometry', *ACM Transactions on Graphics,* 23/2: 97–110.

Kappraff, J. (1991), *Connections: The Geometric Bridge between Art and Science,* New York: McGraw-Hill.

Kappraff, J. (2000), 'A Secret of Ancient Geometry', in C. A. Gorini (ed.), *Geometry at Work: Papers in Applied Geometry, MAA Notes*, 53: 26–36.

Kappraff, J. (2002 and reprint 2003), *Beyond Measure: A Guided Tour through Nature, Myth and Number,* River Edge (USA), London and Singapore: World Scientific.

Klee, P. (1953), *Pedagogical Sketchbook,* W. Gropius and L. Moholy-Nagy (eds), London: Faber and Faber.

Klee, P. (1961), *The Thinking Eye: The Notebooks of Paul Klee,* J. Spiller (ed.) and R. Manheim (trans.), London: Lund Humphries.

Lawlor, R. (1982), *Sacred Geometry: Philosophy and Practice,* London: Thames and Hudson.

Le Corbusier (1954 and 2nd edn. 1961), *The Modular,* trans. P. de Francia and A. Bostock, London and Boston: Faber and Faber.

Lee, A. J. (1987), 'Islamic Star Patterns', *Muqarnas*, 4: 182–97.

Lidwell, W., Holden, K., and Butler, J. (2003), *Universal Principles of Design,* Gloucester: Rockport.

Lipson, H., Pollack, J. B., and Suh, N. P. (2002), 'On the Origin of Modular Variation', *Evolution,* 56/8: 1549–56.

Livio, M. (2002), *The Golden Ratio: The Story of Phi, The World's Most Astounding Number,* New York: Broadway Books.

Lu, P. J., and Steinhardt, P. J. (2007), 'Decagonal and Quasi-crystalline Tilings in Medieval Islamic Architecture', *Science,* 315: 1106–10.

Lupton, E., and Abbot Miller, J. (eds) (1993), *The ABC's of the Bauhaus and Design Theory,* London: Thames and Hudson.

Lupton, E., and Phillips, J. C. (2008), *Graphic Design: The New Basics,* New York: Princeton Architectural Press.

March, L. (2001), 'Palladio's Villa Emo: The Golden Proportion Hypothesis Rebutted', *Nexus Network Journal,* 3/2: 85–104.

Markowsky, G. (1992), 'Misconceptions about the Golden Ratio', *College Mathematics Journal,* 231: 2–19.

Marshall, D.J.P. (2006), 'Origins of an Obsession', *Nexus Network Journal,* 8/1: 53–64.

Meenan, E. B., and Thomas, B. G. (2008), 'Pull-up Patterned Polyhedra: Platonic Solids for the Classroom', in R. Sarhangi and C. Sequin (eds), *Bridges Leeuwarden, Mathematical Connections in Art, Music and Science,* Conference Proceedings, St Albans: Tarquin.

Melchizedek, D. (2000), *The Ancient Secret of the Flower of Life,* vol. 2, Flagstaff, AZ: Light Technology Publishing.

Meyer, F. S. (1894), *Handbook of Ornament: A Grammar of Art, Industrial and Architectural,* 4th ed. New York: Hessling and Spielmayer, reprinted (1957) New York: Dover, and (1987) as *Meyer's Handbook of Ornament,* London: Omega.

Meyer, M., and Seliger, R. (1998), 'Product Platforms in Software Development', *Sloan Management Review,* 40/1: 61–74.

Naylor, G. (1985), *The Bauhaus Reassessed: Sources and Design Theory,* London: Herbert Press.

Necipoglu, G. (1995), *The Topkapi Scroll: Geometry and Ornament in Islamic Architecture,* Santa Monica: Getty Center for the History of Art and the Humanities.

Olsen, S. (2006), *The Golden Section: Nature's Greatest Secret,* Glastonbury, Somerset (UK): Wooden Books.

Ostwald, M. J. (2000), 'Under Siege: The Golden Mean in Architecture', *Nexus Network Journal,* 2: 75–81.

Özdural, A. (2000), 'Mathematics and Arts: Connections between Theory and Practice in the Medieval Islamic World', *Historia Mathematica,* 27: 171–201.

Padwick R., and Walker, T. (1977), *Pattern: Its Structure and Geometry,* Sunderland: Ceolfrith Press, Sunderland Arts Centre.

Pearce, P. (1990), *Structure in Nature Is a Strategy for Design,* Cambridge, MA: MIT Press.

Post, H. (1997), 'Modularity in Product Design Development and Organization: A Case Study of Baan Company', in R. Sanchez and A. Heene (eds), *Strategic Learning and Knowledge Management,* New York: John Wiley & Sons.

Racinet, A. (1873), *Polychromatic Ornament,* London: Henry Sotheran, republished (1988) as *The Encyclopedia of Ornament,* London: Studio Editions.

Reynolds, M. A. (2000), 'The Geometer's Angle: Marriage of Incommensurables', *Nexus Network Journal,* 2: 133–44.

Reynolds, M. A. (2001), 'The Geometer's Angle: An Introduction to the Art and Science of Geometric Analysis', *Nexus Network Journal,* 3/1: 113–21.

Reynolds, M. A. (2002), 'On the Triple Square and the Diagonal of the Golden Section', *Nexus Network Journal,* 4/1: 119–24.

Reynolds, M. A. (2003), 'The Unknown Modular: The "2.058" Rectangle', *Nexus Network Journal,* 5/2: 119–30.

Rooney, J., and Holroyd, F. (1994), *Groups and Geometry: Unit GE6, Three-dimensional Lattices. and Polyhedra,* Milton Keynes: Open University.

Rosen, J. (1989), 'Symmetry at the Foundations of Science', *Computers and Mathematics with Applications,* 17/1–3: 13–15.

Rowland, A. (1997), *Bauhaus Source Book,* London: Grange.

Sanderson, S. W., and Uzumeri, M. (1997), *Managing Product Families,* New York: McGraw-Hill.

Sarhangi, R. (2007), 'Geometric Constructions and Their Arts in Historical Perspective', in R. Sarhangi and J. Barrallo (eds), *Bridges Donastia,* Conference Proceedings (San Sebastian, University of the Basque Country), St Albans: Tarquin.

Schattschneider, D. (2004), *M. C. Escher: Visions of Symmetry,* London: Thames & Hudson.

Schlemmer, O. (1971 ed.), *Man: Teaching Notes from the Bauhaus,* H. Kuchling (ed.), H. M. Wingler (pref.) and J. Seligman (trans.), London: Lund Humphries.

Scivier, J. A., and Hann, M. A. (2000a), 'The Application of Symmetry Principles to the Classification of Fundamental Simple Weaves', *Ars Textrina,* 33: 29–50.

Scivier, J. A., and Hann, M. A. (2000b), 'Layer Symmetry in Woven Textiles', *Ars Textrina,* 34: 81–108.

Seeley, E. L. (trans.) (1957), *Lives of the Artists, by Giorgio Vasari,* New York: Noonday Press.

Senechal, M. (1989). 'Symmetry Revisited', *Computers and Mathematics with Applications,* 17/1–3: 1–12.

Shubnikov, A.V., and Koptsik, V.A. (1974), *Symmetry in Science and Art,* New York: Plenum Press.

Speltz, A. (1915), *Das Farbige Ornament aller Historischen Stile,* Leipzig: A. Schumann's Verlag, and republished (1988) as *The History of Ornament,* New York: Portland House.

Stephenson, C., and Suddards, F. (1897), *A Textbook Dealing with Ornamental Design for Woven Fabrics,* London: Methuen.

Stevens P. S. (1984), *Handbook of Regular Patterns: An Introduction to Symmetry in Two Dimensions,* Cambridge, MA: MIT Press.

Stewart, M. (2009), *Patterns of Eternity: Sacred Geometry and the Starcut Diagram,* Edinburgh: Floris Books.

Sutton, D. (2007), *Islamic Design,* Glastonbury: Wooden Books.

Tennant, R. (2003), 'Islamic Constructions: The Geometry Needed by Craftsmen', *International Joint Conference of ISAMA, the International Society of the Arts, Mathematics and Architecture,* and *BRIDGES, Mathematical Connections in Art, Music and Science,* University of Granada, Spain.

Thomas, B. G. (2012), University of Leeds, Personal communication with the author.

Thomas, B. G., and Hann, M. A. (2007), *Patterns in the Plane and Beyond: Symmetry in Two and Three Dimensions,* Leeds: University of Leeds International Textiles Archive (ULITA) and Leeds Philosophical and Literary Society.

Thomas, B. G., and Hann, M. A. (2008), 'Patterning by Projection: Tiling the Dodecahedron and Other Solids', in R. Sarhangi and C. Sequin (eds), *Bridges Leeuwarden, Mathematical Connections in Art, Music and Science,* Conference Proceedings, St Albans: Tarquin.

Thompson, D. W. (1917, 1961, and 1966), *On Growth and Form,* Cambridge: Cambridge University Press.

Tower, B. S. (1981), *Klee and Kandinsky in Munich and at the Bauhaus,* Ann Arbor, MI: UMI Research Press.

Wade, D. (1976), *Pattern in Islamic Art,* Woodstock: Overlook Press.

Wade, D. (2006), *Symmetry. The Ordering Principle,* Glastonbury: Wooden Books.

Washburn D. K., and Crowe D. W. (1988), *Symmetries of Culture: Theory and Practice of Plane Pattern Analysis,* Seattle and London: University of Washington Press.

Washburn, D. K., and Crowe, D. W. (eds) (2004), *Symmetry Comes of Age: The Role of Pattern in Culture,* Seattle and London: University of Washington Press.

Watts, D. J., and Watts, C. (1986), 'A Roman Apartment Complex', *Scientific American,* 255/6: 132–40.

Weyl, H. (1952), *Symmetry,* Princeton: Princeton University Press.

Williams, K. (2000), 'Spirals and the Rosette in Architectural Ornament', in C. A. Gorini (ed.), *Geometry at Work: Papers in Applied Geometry, MAA Notes,* 53: 3–11.

Williams, R. W. (1972, reprinted 1979), *The Geometrical Foundation of Natural Structures: A Source Book of Designs,* New York: Dover.

Willson, J. (1983), *Mosaic and Tessellated Patterns: How to Create Them,* New York: Dover.

Wilson, E. (1988), *Islamic Designs,* London: British Museum.

Wolchonok, L. (1959), *The Art of Three-Dimensional Design,* New York: Dover.

Wong, W. (1972), *Principles of Two-Dimensional Design,* New York: Van Nostrand Reinhold.

Wong, W. (1977), *Principles of Three-Dimensional Design,* New York: Van Nostrand Reinhold Company.

Woods, H. J. (1936), 'The Geometrical Basis of Pattern Design. Part 4: Counterchange Symmetry in Plane Patterns', *Journal of the Textile Institute: Transactions,* 27: T305–20.

Worren, N., Moore, K., and Cardona, P. (2002), 'Modularity, Strategic Flexibility and Firm Performance: A Study of the Home Appliance Industry', *Strategic Management Journal,* 23/12: 1123–40.

index

402373